Power plants

Manchester University Press

Power plants

Bioenergy, vegetal labour and the politics of productivity

James Palmer

Manchester University Press

Published by Manchester University Press
Oxford Road, Manchester, M13 9PL

www.manchesteruniversitypress.co.uk

British Library Cataloguing-in-Publication Data
A catalogue record for this book is available from the British Library

ISBN 978 1 5261 9212 7 hardback

First published 2026

EU authorised representative for GPSR:
Easy Access System Europe, Mustamäe tee 50,
10621 Tallinn, Estonia
gpsr.requests@easproject.com

Typeset by Newgen

Contents

Figures

Tables

Acknowledgements

Perhaps fittingly for a text advocating a rejection of productivism, much time has been squandered in the slow writing of this book. The idea for *Power Plants* dates to 2018, when I arrived, as a Vice-Chancellor's Fellow in Environmental Social Science and Humanities, at the University of Bristol. The book's roots, however, go further back to research on biofuels policy undertaken at the University of Cambridge between 2008 and 2013. There, I was extremely fortunate to benefit from the intellectual guidance of Professor Susan Owens—a brilliant scholar of the environmental science–policy interface, a paragon of policy influence through her dedicated work as an expert adviser in numerous settings, and above all an inspiring role model for building academic success upon kindness, modesty, a forensic attention to detail and a healthy dose of mischievous humour.

Over the years, many individuals have provided astute feedback and advice on my research and writing, encouraged me to keep going or—through the gift of friendship—simply made coming into work feel much less like work than it probably ought to have done. In Cambridge, I gained much from discussions with Michael Bravo, Bill Adams, Keith Richards and many others, not least through the Travelling Knowledges Reading Group. Robert Doubleday generously involved me in some of the early work of the Cambridge Centre for Science and Policy, granting me privileged access to conversations I had little to contribute to but much to learn from. At Cambridge I also met Eszter Kovacs—not just a razor-sharp critic, but a cherished friend who is unfailingly fair, frank and funny (normally in all the right places).

Later on, at Oxford, I owed much to the generosity of Alasdair Rogers and Richard Washington at Keble College, to the support of Tim Schwanen (at the Transport Studies Unit) and to the wisdom of the late Steve Rayner (at the Institute for Science, Innovation and Society). Being immersed for a short time in the new materialist ambience of Oxford's School of Geography and the Environment—clinging barely, if at all, to the intellectual coattails of Sarah Whatmore and her acolytes—appears in retrospect as an "obligatory passage point" for this book (to call on a certain phrase). This was the milieu which first opened my eyes to the potentials of writing not about biofuels or bioenergy in narrow terms, but about the energies of plants more broadly. No less vitally, Oxford also gave me more lifelong friends—Joe Gerlach, Thomas Jellis and Maan Barua. All have offered more support and inspiration (and no shortage of appropriately lubricated levity) than I can ever repay.

Most recently, to Bristol, where my luck in securing a Vice-Chancellor's Fellowship owes immeasurably to Joe Gerlach again (who kindly tested Bristol's waters for me) and Maria Fannin. The Fellowship itself offered priceless space and time for me to develop new strands of bioenergy-focused research—on wood pellet manufacturing in working forests and the use of genetic engineering in advanced biofuel research—which find their places here in Chapters 3 and 4. Now, as a lecturer in Bristol's School of Geographical Sciences, I incur a near-daily debt to the formidable intellectual talents of my colleagues in the Historical and Cultural Geographies and Political Ecologies Research Groups. Not only are these some of the sanest people left in UK academia in the mid-2020s, but also, more than once, they have sat patiently through unconvincing presentations of early versions of chapters of this book, offering enthusiastic and unerringly incisive feedback.

Along the way, whether about this book in particular or academic life more generally, many others have offered friendly encouragement and advice—thanks here go especially to Mike Hulme, Harriet Bulkeley, Tim Forsyth, Alex Loftus, Susan Parnell, Jonathan Rigg and Rachel Flecker. In the final stages of writing, I was lucky to benefit from stimulating conversations and generous feedback at two wonderful workshops on energy resource-making in Copenhagen—for this, sincere thanks go to organisers Peter Karnøe, José Ossandón and Ask Greve Johansen, as well as to fellow attendees, especially Béatrice Cointe and Gavin Bridge. For shepherding the book from its precarious state as mere words on my screen to a tangible object with the potential now to live a life of its own, I am hugely grateful to the ever-enthusiastic Shannon Kneis and Chia-Chieh Liu at Manchester University Press, and to Cheryl Adam and Jess Farr-Cox for their expert copy-editing and indexing, respectively.

As befits any writing project of this ambition (or at least this gestation period), I am in debt to countless individuals and organisations beyond those named above. This includes colleagues and anonymous reviewers who have shaped earlier iterations of this book's arguments, particularly those set out in Chapter 1 (first trialled in "Rethinking Plant Power: Bioenergy, Vegetal Labor, and Postproductivist Energy Cultures," *Environmental Humanities* 16, no. 3 (2024): 661–679) and Chapter 3 (which builds upon "Putting Forests to Work: Enrolling Vegetal Labor in the Socioecological

Fix of Bioenergy Resource Making," *Annals of the American Association of Geographers* 111, no. 1 (2021): 141–156). A greater share of the thanks must go, however, to the people from so many different walks of life who have given up their time, over the years, to provide their perspectives on modern bioenergy, whether in interviews or more informally in the field. In the end, I can only truly take credit for the ways in which the end result underdelivers against the kind of book which would have done real justice to their rich insights and principled convictions.

The final thanks, however, must go to my family: to my parents, Ron and Donna, and my sister, Hannah, for your enduring support and encouragement, to my partner, Marion, for her unwavering love (and no little critical intellectual input!) and—most recently and miraculously—to our two beautiful daughters, Elsa and Leonie. With a 3-year-old and 1-year-old now to keep up with, life may be hectic, but ultimately this book would never have been finished without the profound happiness and sense of purpose that parenthood has brought with it. So, girls, please don't grow up too fast.

Abbreviations

BECCS	Bioenergy with carbon capture and storage
BECCUS	Bioenergy with carbon capture, utilisation and storage
BECS	Biomass energy with carbon sequestration
BRC	Bioenergy Research Centers (of the US Department of Energy)
CABBI	Center for Advanced Bioenergy and Bioproducts Innovation
CAP	Common Agricultural Policy (of the European Union)
DfT	Department for Transport (of the United Kingdom)
DOE	Department of Energy (of the United States)
EU	European Union
FAO	Food and Agricultural Organization (of the United Nations)
GHG	Greenhouse gas
IEA	International Energy Association
IIASA	International Institute for Applied Systems Analysis
ILUC	Indirect land-use change
IPCC	Intergovernmental Panel on Climate Change
IRENA	International Renewable Energy Association
JRC	Joint Research Centre (of the European Commission)
OECD	Organisation for Economic Co-operation and Development
SAF	Sustainable aviation fuel
TAG	Triacylglycerol
UNFCCC	United Nations Framework Convention on Climate Change
WEF	World Economic Forum
WTO	World Trade Organization

Introduction: power plants—bringing energy to life

> The formation of a vegetable is a process of winding up, while the formation of an animal is a process of running down.
>
> —John Tyndall, *Heat Considered as a Mode of Motion*, 1869

> The challenge now is ... to turn our planet from something with a past to mine to something with an insolation to exploit.
>
> —Oliver Morton, *Eating the Sun*, 2007

Twenty-first-century energy systems are increasingly animated by living plants. In the European Union (EU) and the United States (US), more than 5 per cent of standard petrol and diesel is now derived not from fossil oil deposits but from agricultural crops such as wheat, sugarcane, oilseed rape, palm oil and soy. Globally, liquid transport biofuel production totalled more than 150 billion litres in 2023.[1] While existing government plans are already forecast to nearly treble this figure by 2050, the International Renewable Energy Agency (IRENA) advocates a greater than four-fold increase, to more than 650 billion litres, to help the transport sector contribute fully to restricting global temperature increases this century to below two degrees Celsius.[2] In the electricity sector meanwhile, numerous countries across Europe and beyond, including Japan and South Korea, are converting coal-fired power generation facilities to run instead on solid biomass fuels, typically in the form of compressed wood pellets.[3] In the United Kingdom (UK), millions of tonnes of these pellets are already imported each year, derived from trees growing in the forests of not only neighbouring European countries like Latvia and Portugal, but also Canada and—most prominently—the US. A recent industry declaration argues that this form of "sustainable wood-based bioenergy will expand to deliver around 20 EJ [exajoules] by 2050, up from 7 EJ today".[4] No less ambitious, meanwhile, are plans to replace vast quantities of conventional jet fuel with so-called "sustainable" aviation biofuels in the coming decades. In the US, for example, hundreds of millions of dollars of government funding have been invested in specialist bioenergy research centres where plant scientists and molecular biologists strive to develop new, genetically engineered varieties of crops conventionally grown for sugar or starch—such as sugarcane and sorghum—so that they might yield large volumes of oil for potential use as an alternative jet fuel.[5] Finally, and perhaps most ambitiously of all, in elite scientific and political deliberations about the threat of global climate change, the photosynthetic activity of plants is routinely earmarked, on a massive scale, as the basis for achieving vast quantities of so-called "negative" greenhouse gas (GHG) emissions, using bioenergy with carbon capture and storage (BECCS) technology to actively remove carbon dioxide from the atmosphere. According to the Intergovernmental Panel on Climate Change (IPCC), limiting global warming to no more than 1.5 degrees above preindustrial levels would require somewhere between 150 and 1,200 billion tonnes of carbon dioxide to be removed from the atmosphere using BECCS before the end of this century.[6]

Whether for the generation of electricity, the provision of heat or the movement of people and goods, diverse efforts to identify future sources of fuel are thus converging upon the dynamic exertions of photosynthesising lifeforms—in short, upon bioenergy—as a large-scale, modern alternative to subterranean reserves of coal, oil and gas. At a global scale, bioenergy accounts for more power today than hydroelectricity, wind, solar and all other renewables combined—making up 70 per cent of the world's renewable energy supply.[7] Yet it also remains "the overlooked giant" of the renewable energy landscape.[8] In fact, when set against more iconic renewable energy infrastructures—not least the unmistakable physical forms taken by large-scale hydropower facilities or extensive arrays of solar panels and wind turbines—recent moves to harness more power from vegetal lifeforms can struggle to appear modern at all.[9] Certainly, the combustion of biomass has only very rarely been subordinate, throughout human civilisation, to other forms of energy generation. Indeed, even in an era of ostensibly decadent fossil fuel-powered industrial capitalism, bioenergy has remained a linchpin for meeting the basic subsistence needs of much of the global population.[10] Modern bioenergy is nonetheless radically different from its more "mundane", preindustrial equivalents, and not only because of its exploding scale and deepening technological sophistication.[11] In fact, this book's core argument is that modern bioenergy presents an opportunity not just to reshape global energy systems but also to reconfigure global energy–society relations, with implications stretching far beyond questions about the environmental or climate impacts of humanity's apparently insatiable demand for fuel. Ultimately, the most radical political potentials of an ongoing modern "bioenergy boom"[12] can be found, I argue, in the potential for plants to help us rethink deep-seated commitments to efficiency, productivity and growth as the principal hallmarks of virtuous energy use.

While the modern bioenergy revolution may be easy to overlook, it is already vital to the day-to-day functioning of many of the world's most affluent economies. Intensifying efforts to put plants to productive work as renewable fuels, however, are not held together by a single, centrally coordinated programme of planned energy transition. Diverse socioeconomic and political drivers are at play in each case, cutting across multiple scales and policy jurisdictions. Even so, assessed at a global scale, existing governmental commitments are expected to see modern bioenergy's proportional contribution to total worldwide primary energy supply double by the middle of this century (from 5 to 10 per cent), against a more modest background increase in overall primary energy demand of just under 20 per cent.[13] While the various policies that account for this growth are diverse, at a discursive level they are frequently justified by one or more of just a handful of key recurring arguments. To begin with, political motivations for ratcheting up modern bioenergy production have often been linked to the imperative of enhancing the security and resilience of strategically important national or regional energy systems, whether by reducing their reliance upon imported (usually fossil) fuels, or more simply by diversifying the mix of resources upon which extant electricity, heating and transport infrastructures depend. At another level, meanwhile, the use of bioenergy—and particularly of so-called "advanced" biofuels, whose efficiency as energy carriers is enhanced through biotechnology—is valorised for its capacity to contribute to a wider political agenda: that of harnessing the intellectual and technical capital of the biological sciences within a newly reformed, "sustainable and circular" bioeconomy.[14] According to this agenda, the bioeconomy will supposedly resolve contradictions between commitments to continuous prosperity and growth on the one hand, and the imperatives of social inclusivity, environmental protection and climate neutrality on the other.[15] More straightforward than either of these first two sets of ideas, however, is a third view of modern bioenergy as a vital tool—at least in theory—for tackling anthropogenic climate change, ostensibly by reducing the amount of carbon dioxide that is added to the atmosphere by global energy consumption.[16] In this latter respect, plants are increasingly coming to be viewed not only as a means of production for energy, but as a vital "means of climate production" too.[17]

Viewed through the lens of climate change, modern bioenergy might reasonably be classified as just one among many components of what geographers Matt Huber and James McCarthy have described as a broader, ongoing "return to a surface energy regime".[18] Here the combustion of biomass takes its place within a wider roster of renewable energy technologies whose collective purpose is to eradicate society's dependence upon fossil fuels by instead tapping into the "more immediate circulations of solar energy" that are continuously available (in one form or another) above ground.[19] What matters most about bioenergy, from this perspective, is its claimed ability—much like solar, wind and hydropower—to sustain the frenetic pace of contemporary human social and economic activity without threatening the stability of the climatic and environmental conditions within which that activity plays out. Accordingly, it can often seem as though the most important questions to ask about modern bioenergy are questions about the efficacy of its contributions to slowing the pace of anthropogenic climate change—questions that demand an interrogation of the patterns of land-use change associated with different kinds of biofuel production, of the efficiency of the techniques used to process trees and crops into combustible hydrocarbon fuel and of the geographies of the supply chains through which various biomass feedstocks are linked up to the needs of disparate energy markets and end users. When combined with accounts of the social and environmental impacts of bioenergy production—for instance upon biodiversity, water quality, food security or land rights—such assessments aim to furnish policymakers with a reliable basis for discriminating between apparently "good" biofuels and "bad" biofuels, and ultimately for optimising and refining the precise role of bioenergy within a wider mix of renewables collectively charged with bringing climate stabilisation into being.

To see bioenergy exclusively in such technoscientific terms is, however, to overlook its most fundamental distinguishing characteristic as a renewable energy resource. Put simply, unlike other renewables, bioenergy manifests not through abiotic processes—whether of atmospheric turbulence, hydrological flows or simply solar insolation itself—but rather as the result of the active metabolism of living organisms. Bioenergy, in other words, is energy that derives directly from life. Plants, as autotrophs, are not freely occurring, but must instead "continually remake themselves out of simple substances that are present in their surroundings".[20] The metabolic activities that facilitate that remaking—photosynthesis and biomass production—are moreover the basis not only for the continuous (re)production of the terrestrial biosphere, but also eventually for the formation of fossil fuel deposits themselves. Indeed, if fossil fuels like coal can best be described—in the words of William Stanley Jevons—as "a spring, wound up during geological ages for us to let down", it is only thanks to the metabolic capabilities of plants that the spring could ever have been tightened in the first place.[21] Consequently, even as modern bioenergy forms part of a wider "return to the surface" for energy in the twenty-first century,[22] its relationship with the subterranean energy regime of the fossil fuel era is far more intimate than its relationship with other renewables. Indeed, in some contexts, increased bioenergy use appears actively to forestall any potential "transition" away from fossil fuels by requiring those fuels to continue to be burned, as when liquid biofuels are blended at low concentrations into conventional petrol and diesel, for example.[23] Even if such interdependencies were to be eradicated, moreover, some contend that any challenge posed by bioenergy to coal, oil or gas remains "purely notional",[24] since it replaces fossilised hydrocarbons only with biologically derived ones, thereby failing to address the deeper dependency of contemporary energy systems upon combustible fuels derived ultimately from photosynthesis.[25]

Against this backdrop, my aim in this book is to showcase just how much is really at stake in the growing enrolment into contemporary energy systems no longer just of the fossilised remnants of prehistoric vegetal life, but also of the ongoing, real-time exertions of living plants. In pursuing this objective, the book will explore not merely how bioenergy derived from plants is serving to displace a steadily increasing volume of energy previously derived from fossil fuels, but also how the turn to large-scale, modern bioenergy is facilitating a broader restructuring of contemporary energy–society relationships along geographical, material and political–economic lines. More specifically, the book will explore how the production of modern bioenergy is bringing into being new geographical orderings of resource provisioning, new kinds of vegetal natures and new sources of economic value and capital accumulation. At the same time, however, the book also aims to offer more critical reflection on the potential of plants to help open up opportunities for society to think differently about how it relates to energy itself. In other words, the book also asks whether plants, as living organisms which remake both themselves and their environments as they grow, might potentially have more to teach us about the roles that energy could play—in shaping desirable post-fossil fuel societies and economies—than we have hitherto been open to noticing.

Modern bioenergy: from governance object to governance logic

Much academic research approaches modern bioenergy as a discrete technoscientific object needing to be mobilised as efficiently and sustainably as possible within contemporary energy and climate governance. The key purposes of bioenergy, from this standpoint, are firstly to help hasten a wider transition away from fossil fuels and towards renewable energy, and secondly to stabilise the global climate system such that dangerous and potentially irreversible climate warming is avoided. This is the view which informs work assessing the thermodynamic efficiency or GHG "footprint" of different kinds of biofuel production, for example, as well as research examining the broader social and environmental impacts of specific bioenergy supply chains. But it is also often detectable in more critical academic research analysing the governance tools and discourses used to legitimise and incentivise bioenergy production in different political contexts. In fact, across these diverse fields of scholarship, whether bioenergy is greeted with enthusiasm or scepticism, it is almost always understood as an external input designed to sustain, rather than unsettle or reshape, existing social and economic systems. In this book, by contrast, I approach modern bioenergy not as a discrete object of energy or climate governance regimes, but rather as a governance logic in its own right. From this perspective, bioenergy cannot be understood merely as a renewable resource input, for which demand emerges only as the byproduct of wider social, economic or environmental imperatives. Rather, modern bioenergy—I argue—constitutes a set of guiding principles and ideas for working with plants to create new kinds of energy systems which, despite their many differences from fossil fuel-powered ones, can nonetheless reinforce the positive economic and cultural associations of large-scale energy use itself. The key idea here, in other words, is that modern bioenergy does more than simply *fuel* ongoing human habits of intensive socioeconomic activity. More fundamentally, it also aims to prefigure the continued intensive consumption of energy as a desirable, and indeed necessary, preoccupation for post-fossil fuel societies in the first place.

By adopting this overarching approach, this book does not intend to depict modern bioenergy as a single, coherent political project being advanced systematically and uniformly across all parts of the globe through a coordinated set of governmental, scientific and commercial activities. Nor does it regard the outcomes of an ongoing expansion of modern bioenergy systems as necessarily predetermined or inevitable. Present-day efforts to derive large volumes of renewable energy from plants are manifold, not only in their material substance and geographical configuration, but also in respect of the ways in which they emerge from situated and historically mediated interactions between scientific evidence, commercial interests, political cultures and discursive framings of energy's position in relation to broader societal and environmental challenges. Accordingly, my argument here is that there is far more at stake in modern bioenergy projects than the simple substitution of fossil fuels, which are effectively "buried sunshine" originating in "deep time",[26] with what might be thought of as living fuels predicated upon solar insolation arriving at the earth's surface in real time. Certainly a large part of bioenergy's appeal—evident across diverse political and geographical contexts—derives from its potential, at least in theory, to render the burning of hydrocarbons environmentally benign. But to focus solely on bioenergy's environmental credentials is to overlook the significant ways in which large-scale industrial energy systems have historically reshaped not only the composition of the earth's atmosphere, but also the geopolitical terrain and cultural values of global capitalist society itself.[27] The specific material properties and affordances of fossil fuels, for example, have been identified as crucial in facilitating what anthropologist Dominic Boyer has termed regimes of "energopower" characterised by centralised political control over the terms of resource extraction, distribution and use.[28] Moreover, as energy historian Cara Daggett has recently expounded, fossil fuels have also served to naturalise fundamental notions of industriousness, productivity and indeed work itself as virtuous human characteristics.[29] By contrast, while living plants have been used to derive mundane forms of bioenergy for millennia, it is only very recently that they have begun to act as a significant ecological foundation not just for industrial energy systems, but also for shoring up prevailing attitudes towards economic growth and human progress writ large.[30]

One of the central aims of this book is therefore to foreground the discursive, scientific, commercial and political work required to govern vegetal lifeforms—and the wider socioecological relations within which those lifeforms exist—in ways conducive to the production of energy resources whose cultural and political–economic affordances, and not merely material properties, can effectively mimic those of fossil fuels. In pursuing this aim, moreover, an account will be offered of modern bioenergy as a fluid and intrinsically open-ended governance apparatus, one with the potential always to be reinterpreted, reconfigured and resisted.[31] Following Michel Foucault, an apparatus can be understood as a "system of relations" through which a "thoroughly heterogeneous ensemble [of elements] consisting of discourses, institutions, architectural forms, regulatory decisions, laws, administrative measures, scientific statements, [and] philosophical, moral and philanthropic propositions" are held together.[32] In the case of modern bioenergy, I argue that the dominant strategic function of this system of relations is to promote the existence of vegetal lifeforms capable of taking on the work (or labour[33]) previously performed by fossil fuels not just as energy carriers but also as sources of productive political power, with stark implications for the geographies, materialities and economies of energy generation more broadly. More specifically, this means that I analyse modern bioenergy as an apparatus through which new configurations of control and authority are being established over the ownership and use of land, over the production of new kinds of nature and economic value (linked not just to energy use but also to climate stabilisation) and ultimately over the terms on which both individual identities and broader societal ideals continue to be shaped by relations with energy in post-fossil fuel societies.

In seeking to substantiate these overarching arguments, the core of the book will examine the emergence, consolidation and effects of the modern bioenergy apparatus as it has manifested in several distinct cases, focusing on developments taking place in Anglo-American and European contexts since the turn of the twenty-first century. At one level, these cases will be used to explicate how modern bioenergy brings about new geographies of energy production and consumption that not only reshape rural landscapes and restructure rural–urban relationships, but also reconfigure international energy geographies, with significant consequences for the colonial and justice dimensions of contemporary resource geopolitics.[34] At another level, meanwhile, the analysis will focus on modern bioenergy's propensity to bring into being new energy materialities—including in the form of bespoke, dedicated energy crops enhanced through biotechnology—and reflect on the implications of these innovations for societal understandings of the relationships between liveliness and productivity, growth and labour.[35] Finally, a third set of analytical preoccupations will address the distinctiveness of modern bioenergy's political–economic logics and their underlying conceptualisations of value, viewing bioenergy as a key component of efforts to kickstart wider so-called "bioeconomies".[36] In this last respect, attention will centre on the capacities of vegetal lifeforms not only to capture and convert solar energy into a useful commodity as a combustible energy carrier, but also—in effecting these metabolic transformations—to help regulate the composition of the global atmosphere itself.

While the scientific, commercial and sociopolitical activities which animate each of the cases explored in the book are confined to Anglo-American and European contexts, the geographies of the bioenergy production systems and supply chains which are being envisaged and brought into being in each case are invariably more extensive and cosmopolitan in scope. Indeed, the production of uneven and in many cases distinctly neocolonial energy geographies—linking generally more affluent, bioenergy-consuming regions in the global North with generally poorer, bioenergy-producing regions disproportionately located in the global South—is a recurring hallmark of today's modern bioenergy apparatus. Nonetheless, this book's focus on initiatives emerging predominantly from Anglophone and northern European political contexts means that its assessment cannot claim to be exhaustive. Rather, the examples that I engage with are intended to illustrate the wide diversity of available pathways—never entirely smooth or uncontested, and some already more fully realised than others—through which efforts to put vegetal life to productive work as modern bioenergy in the coming decades may proceed and expand. From a theoretical perspective, engagement with these cases is intended to help explicate the relevance and implications of modern bioenergy for several broader, ongoing debates involving critical geographers alongside anthropologists, political economists and scholars of science and technology studies and the environmental humanities more broadly. As the next section outlines, these debates transcend the specific realms of energy transition and indeed climate governance, cutting to the heart of more fundamental categorical distinctions between life and labour, energy and work. Ultimately, these debates raise questions about the underlying meanings and purposes of efficiency, productivity, growth and energy in post-fossil fuel societies—questions towards which a critical analysis of modern bioenergy promises to add novel insights and provocations.

Putting down roots: modern bioenergy and the politics of vitality and value

On the surface, this is a book concerned with one specific element of an ongoing renewable energy transition. While there is a significant and important body of critical social scientific work that has sought to assess the environmental and socioeconomic impacts of different kinds of bioenergy production, the core aim of this work is instead to illuminate how vegetal lifeforms, and more specifically photosynthetic metabolism, are being enrolled into a broader restructuring of global energy–society relations in the twenty-first century. Bioenergy may appear to constitute a straightforward substitute for fossil fuels, but the work of successfully replacing coal, oil and gas with alternatives derived from living plants involves more than the mere practicalities of resource procurement. Deeper ideas about the ultimate purposes of vitality and value, and the relationships that should exist between them, are also at stake. By approaching modern bioenergy as a regulatory logic, this book conceives of efforts to derive fuel from vegetal life as themselves productive of new geographies, materialities and political economies of energy. In so doing, its analysis will draw upon, and seek to advance, several broader and interrelated scholarly debates, all sharing a concern with ongoing transformations to the relationships linking vitality—whether human, vegetal or otherwise—with value in contemporary capitalist economies.

Ecological Marxism and energopolitics

Critical geographers and political ecologists have assembled rich conceptual frameworks for understanding relationships between the productive forces and social interactions which animate capitalist accumulation on the one hand, and the realms of nonhuman nature and the environment on the other. Central to much of this work has been an ecological Marxist view of environmental degradation—in line with James O'Connor's definition of the "second contradiction" of capitalism—as innate to capitalist production itself; put simply, capital is confounded by its tendency routinely to "underproduce" the environmental conditions necessary for continued production to be feasible.[37] For geographers and others drawing on this perspective, however, the second contradiction of capitalism does not lead directly to a single, definitive ecological crisis. Instead, it is the basis for ongoing political struggles around what Scott Prudham has termed the "social (re)production of new natures, both for the purposes of renewed accumulation and for broader societal goals".[38] Accordingly, much critical geographical work, particularly since the turn of the century, has sought to unpack the processes through which various fractions of capital—whether within or beyond the energy sector—have deliberately transformed the biophysical world in order to achieve so-called "socioecological fixes" to crises of accumulation brought about by environmental degradation.[39] Such work has charted the emergence of new accumulation strategies in water governance, forestry, wetland restoration and fisheries management, ranging from the "real subsumption" of nature to capital (whereby the speed and direction of biophysical and ecological processes are optimised to enhance value production[40]) through to the development of overtly neoliberal policy frameworks for commodifying processes of ecological conservation, restoration and repair themselves.[41] From the perspective of these literatures, nature must ultimately be understood not only as something that is produced through capital, but also as an accumulation strategy in its own right.[42]

Despite their richness, however, critical accounts of the relationships between capitalism and nature have sometimes been rebuked for casting ecological forces into a restricted roster of agentive positions. Indeed, it can sometimes seem that nature can only manifest either as a commodified output of capitalist production or else as a set of external, uncommodified conditions within which production must take place. Yet for scholars including Timothy Mitchell and Matthew Huber, the material ecologies of industrial energy systems also serve to shape more everyday practices, beyond the formal spheres of production and consumption, that lend meaning and legitimacy to capitalism as a set of social relations, and not merely as an economic system.[43] In his work focusing on oil in the twentieth-century US, for instance, Huber pinpoints "the incredible reservoir of 'work' provided by fossil energy" as an indispensable ecological foundation not just for intensive industrial regimes of commodity production, but also for tenacious ideas about the appeal of an "American way of life" structured around fundamentally capitalist precepts—including privatisation, competition and entrepreneurialism.[44] Oil's material ecologies have been indispensable, in this account, to the emergence of suburban geographies of social reproduction centred around single-family homes, automobility and—ultimately—a view of life itself as a coherent, privatised space of freedom "entirely produced by and reducible to one's own ... entrepreneurial efforts".[45] From this vantage point, the ecological foundations of energy systems can be said to enable particular regimes of energopower,[46] in which practices of material production, exchange and accumulation co-evolve with deeper systems of beliefs about how to succeed and get the most out of life more broadly. These ideas encourage us to view the present-day expansion of modern bioenergy, along with other forms of renewable energy, as concerned with more than simply averting an impending environmental crisis. Accordingly, as plants come to fulfil a greater proportion of our demand for energy, one of the central aims of this book is to examine the implications that new vegetal energy ecologies might have for prevailing ideas about the desirability of those ways of life which currently make intensive energy consumption appear necessary and virtuous in the first place.

Working natures and hybrid resource-making

Critical geographers and political economists have long been preoccupied by a need to account for what Margaret Fitzsimmons pithily termed "the matter of nature",[47] not least in work examining the recalcitrant and—to quote Karen Bakker—"uncooperative" tendencies of those elements of the biophysical world required to serve as resource inputs for basic commodity production in realms such as agriculture and water governance.[48] Of late, however, the lively potentials of the nonhuman world have increasingly come to feature in accounts of the forces animating what Marx famously termed the "hidden abode of production" itself.[49] Indeed, building on a broader post-humanist turn evident in the social sciences from the mid-2000s onwards, many recent contributions have embraced an unequivocally relational view of the economy as a realm "constituted from the outset through a set of practices ... including the 'performances' of non-humans".[50] Studies of natural resource extraction in geography and anthropology, for example, have increasingly abandoned a view of the nonhuman realm as a "passive terrain carved and channelled by social agents", in favour of one that sees resources as continually made and remade through a combination of human labour and the "generative capacities of biophysical processes".[51] In studies of economies centred around animal life, meanwhile—whether in work attending to the place of broiler chickens in meat production, Asiatic lions in conservation tourism or parrots and pythons in the exotic wildlife trade—value production is increasingly seen as actively shaped not just by the generative capacities and "liveliness" of nonhumans, but also by what Alyssa Battistoni has more assertively described as "the work of nature".[52]

Taking inspiration originally from Donna Haraway's provocative reinterpretation of Marx's labour theory of value,[53] this seam of scholarship has unsettled the apparent "self-evidence" of basic political–economic categories, not least by postulating "nonhuman labour" as a central component of contemporary systems of capitalist production and exchange.[54]

In contrast to conventional ecological Marxist perspectives, these more recent interventions would view the *first* contradiction of capitalism—its tendency to despoil the productive forces of labour itself—as no less ecological than the second.[55] Accordingly, a view of nonhuman entities as sources of labour power in their own right is now increasingly being extended beyond the animal realm, for instance in work exploring the centrality of soil biota's "microbial labour" to emergent regimes of so-called "ecological intensification" in agriculture, and in studies of the work performed by plants in diverse economic contexts linked to horticulture, viticulture and even urban climate adaptation.[56] The applicability of these ideas to the modern bioenergy apparatus scrutinised in this book is not difficult to discern; bioenergy feedstocks are fundamentally made through the photosynthetic exertions of plants, enticing us to view bioenergy ultimately as the product of not only human work but also what can be termed "vegetal labour".[57] Despite rich veins of scholarship examining industrial production in the agricultural, forestry and energy sectors, however, critical accounts of the relationships between capitalism and nature have only rarely engaged with the one sector—bioenergy—that brings all three of these fields of activity—agriculture, forestry and energy—together. A core concern of this book, therefore, will be to address this gap by examining exactly how the lively potentials of plants—to photosynthesise, grow and sequester carbon dioxide—are being enrolled as sources of labour power in the production of economic value in modern bioenergy regimes, both to generate fuel and to actively reshape the composition of the global atmosphere.

Plants, value and post-work energy futures

While the idea of nonhuman labour offers fertile ground from which to mount critiques of exploitative capitalist production regimes, it remains critical to tread carefully when equating the lively, reproductive capacities of plants, microbes or animals directly with either work or value. To begin with, as scholars in science and technology studies have pointed out, to view "the intrinsic generativity of living biology" as the "ultimate use value and source of all wealth" is to overlook the need for that generativity first to be recognised and rendered measurable, demanding the application of specific technoscientific knowledge frameworks and calculative practices.[58] More fundamentally still, to equate nonhuman life straightforwardly with work is to gloss over the foundational feminist Marxist insight that those activities which come to count as genuine forms of work are—in any society—never self-determining, but established always on the basis of political struggle.[59] While efforts to forge linkages between vitality and economic value must thus be seen as inherently contestable, where modern bioenergy economies specifically are concerned, the very nature of plants can sometimes threaten to put these politically charged questions in the shade. Even the most basic of political–economic distinctions—that which insists on a strict divide between production and reproduction—begins to appear tenuous when one considers that the photosynthetic exertions of plant life serve principally to produce the matter of plants themselves.[60] Moreover, the apparent inability of vegetal lifeforms to refuse to respond to external stimuli not only makes it difficult to distinguish between biological life and productive work, but also plays into a view of the products of vegetal metabolism as "free gifts" of nature.[61] From this vantage point, it can seem difficult—perhaps even futile—to oppose a view of a world playing host to more efficient and productive forms of plant growth, and hence to more abundant bioenergy resources, as anything other than benign and welcome.

But, as anthropologists Sarah Besky and Alex Blanchette have recently pointed out, to contend that "all nonhumans naturally work risks the projection of capital's fixation on the value of human labor onto all of the planet's energies".[62] To fall into this trap is to presume that hard work alone can absolve the planet from an impending ecological crisis—in other words, to believe that the contradictions between capitalism and nature can be resolved simply by putting an ever-increasing proportion of the biophysical world to more efficient and productive work. Such is the vision espoused by protagonists of ecological modernisation and indeed of a sustainable and circular bioeconomy—a stable global climate system and a healthy, biodiverse environment salvaged not only by hard work, but also ultimately *for* it.[63] Beguiling as this view of nature as an untapped reservoir of work may seem, however, the dependence of expanding modern bioenergy systems upon living plants may yet provide opportunities for imagining new kinds of relationships between energy and life, and indeed work and value. As cultural geographers and anthropologists have argued, efforts to govern plant life are intimately bound up with efforts to govern human communities.[64] Yet, as anthropologist Natasha Myers points out, plants' "wily ways also allow them to defy total governance";[65] they frequently subvert human designs and visions, entangling themselves—and others around them—in unexpected new relationships and modes of existence. Even as efforts to expand modern bioenergy systems are often rationalised by a view of biological life as inherently regenerative and therefore hard-working, the realities of situated human–plant relationships, both historical and emergent, might nonetheless serve to call into question a view of energy consumption and productive work as essential components of a life well lived.[66] Today's prevailing regulatory logic of modern bioenergy, in short, not only may encounter resistance from alternative perspectives, but also could serve to catalyse new, more creative modes of thinking about shared future projects for plants and humans to collectively undertake, beyond simply fuelling the expansion of already-existing modes of social and economic life.

An outline of the book

This book represents the culmination of more than a decade of research on bioenergy politics. The empirical data and materials undergirding the core arguments set out in the chapters to come have been derived from site visits to active bioenergy production and research facilities on both sides of the Atlantic, from extensive documentary analysis and from close to one hundred interviews conducted with actors involved in debates about bioenergy production and governance in the UK, EU and US, over a period dating back to 2009. Among those interviewed over this period have been policymakers and their formally appointed advisers, representatives of corporations and other industry bodies active in diverse segments of the bioenergy sector, independent scientific experts and other academic researchers, and officials representing civil society organisations and environmental campaign groups of various stripes. As a critical social scientist, my efforts to define my own relationship to bioenergy throughout this time were fraught with tension and ambiguity. Merely stating that I wish to better understand my interviewees' own perspectives, motivations and concerns about bioenergy rarely proved sufficient to earn their trust; I too have been expected to commit to a point of view, to express enthusiasm or dismay about specific industry initiatives or policy developments and ultimately to demonstrate that mine is a sympathetic as well as an understanding ear. At the same time, however, no amount of polite diplomacy has been able to negate my status as an outsider in these debates, or to disguise my reluctance to pledge unwavering allegiance to any single standpoint or course of action. On more than one occasion, my insistence on the absence of straightforward or definitive answers about bioenergy, and on the need to think more carefully about its implications for society, simply rang hollow or seemed crass—not least when cast against the incontrovertible realities of landscape transformation and ecological deterioration that undoubtedly are being driven by biofuel production in many places.

Yet, despite standing so conspicuously aloof, there have still been significant shifts in my own relationship to bioenergy over the period of this research. As a graduate student, my principal interest in biofuels was as a technoscientific object at the centre of energy and environmental policymaking controversies in which diverse knowledge claims and forms of expertise were being brought into discursive conflict. Over time, however, my interest in the puzzles of policymaking began to give way to questions about the forms of knowledge and power bound up in the making of bioenergy resources themselves. The category of bio*fuels*—through its implied focus on the circulation, consumption and regulation of finished, combustible energy carriers whose purpose is almost always to act as indistinguishable substitutes for fuels of other kinds—began to feel restrictive.[67] The category of bio*energy*, by contrast, allows for the differences between biofuels and other energy resources—most obviously, their inextricable dependence upon living, photosynthesising organisms—to be taken more seriously as a starting point for critical analysis. From this perspective, bioenergy governance must be viewed as more than the mere development of policies aiming to regulate the social and environmental impacts of biofuel production, for instance through sustainability criteria or other supply chain safeguards. Instead, more fundamentally, it is a sphere of activity concerned with the deliberate, industrial-scale extraction of economic value from the metabolic activities of plants themselves. The promotion of modern bioenergy comes to be viewed, in other words, as a global-scale project concerned with harnessing the work not of fossilised organic matter, but of an ever-increasing proportion of the living vegetal kingdom with which we share the earth's surface today.

Accordingly, in Chapter 1, I set out the book's foundational theoretical argument—drawing on insights from resource geography, the energy humanities and especially post-humanist accounts of capitalist production—for recognising and viewing modern bioenergy explicitly as the product of the "work that plants do" when growing and photosynthesising.[68] By focusing on the temporalities of resource formation, this opening chapter unsettles straightforward characterisations of bioenergy as renewable and fossil fuels as exhaustible. Indeed, by scrutinising tensions between the roles that are played—in favourable accounts of modern bioenergy—by the vitality of vegetal lifeforms on the one hand and the productivity of those lifeforms on the other, the chapter firmly rejects views of plants as passive resource inputs within emergent political economies of modern bioenergy production today. Instead, it advocates the concept of "vegetal labour" as a necessary conceptual entry point not just for analysing but also for critiquing the underlying, productivist logic which serves to legitimise the pursuit of modern bioenergy writ large.

With these theoretical foundations in place, the subsequent four chapters all then deal in detail with a particular manifestation of today's modern bioenergy apparatus as it has played out in specific contexts over the past two decades. Each of these chapters engages with a distinct cluster of scientific, commercial and sociopolitical activities that can be viewed as sharing a collective commitment not only to the fulfilment of formal regulatory objectives, but also to the realisation of a broader vision in which modern bioenergy serves to reinforce prevailing, growth-oriented energy cultures of the fossil fuel era. Chapter 2 begins by examining efforts to develop and deploy crop-based liquid biofuels as replacements for petrol and diesel in the European and US road transport sectors, beginning from the early 2000s onwards. Viewing this case as a harbinger of a later, more fully developed regulatory logic of modern bioenergy, the chapter aims to show how an expansion in the use of transport biofuels proceeded hand-in-hand with the production of new energy geographies, undergirded by a view of land itself as first and foremost a blank surface upon which to exploit solar insolation by enacting photosynthesis (linked to questionable discourses of land as degraded, marginal or simply empty). This initial "new rush" for biofuel was much less successful, however, at producing new vegetal natures, and concerned itself hardly at all with the production of new kinds of economic value.

Subsequently, Chapter 3 examines how ideas about work and productivity have shaped the development, particularly since 2010, of a new forest-based bioenergy industry in the US South. Drawing on interviews, site visits and the analysis of documentary data, the chapter argues that biomass wood pellet manufacturing in the region is ultimately predicated upon the coordinated endeavour of both human and vegetal labourers, and that the optimisation of these collaborative working arrangements is central to generating new forms of economic value from forests, linked not strictly to energy production but also to climate change mitigation.

In Chapter 4, the book turns its attention to ongoing efforts on the part of scientific researchers and commercial enterprises not to put existing vegetal lifeforms to work for energy, but rather to actively redesign those lifeforms—through genetic engineering—such that plant growth becomes increasingly indistinguishable from the production and indeed refinement of fuel itself. The chapter will show how imaginaries of a future economy fuelled by abundant, fast-growing, genetically engineered crops serve to depoliticise productivist logics of energy production and use, and their implicit commitments to unrestricted growth, ultimately by producing new kinds of vegetal nature which enact productivism at the molecular level.

Finally, to round off the four cases, Chapter 5 delves into the recent emergence of the technoscientific object of bioenergy with carbon capture and storage (BECCS), upon which hopes that humanity can achieve significant negative GHG emissions this century are today being routinely and widely pinned. The chapter will argue that BECCS constitutes a precarious regulatory project seeking to harmonise the production of new kinds of vegetal natures, new energy geographies *and* new forms of economic value linked directly to climate mitigation, all in the name of reconciling a global ecological imperative for climate stability with an apparently unquestioned global economic imperative for untrammelled growth and limitless energetic expansion.[69]

By examining each of these four cases, the core of the book offers a detailed analysis of the productive political power of modern bioenergy, understood as an apparatus restructuring energy–society relations in post-fossil fuel societies along three broad dimensions: geographical, material and political–economic. While the overall account that emerges from these chapters is highly critical of the societal and environmental implications of modern bioenergy as it is currently envisioned and enacted in affluent societies of the global North, however, it also retains a grain of optimism about the possibilities for bioenergy—through its very dependence upon vegetal life—to provoke those same societies to revisit and rethink ideas about the meanings and purposes of energy itself. Accordingly, in the concluding chapter, the book turns explicitly to the question of how prevailing societal attitudes towards energy might stand to be altered and enriched by a closer attunement to the materialities, temporalities and rhythms of vegetal life. After all, even as dominant approaches to ramping up modern bioenergy tend to view "nature as a vast machine capable of producing mechanical work", the fact remains that fields of crops and stands of trees—even those cultivated under intensive capitalist production regimes—are, for the bulk of their lives, "quiet monuments to nature's power to make men wait".[70] Much as we might pressure vegetal lifeforms to rapidly replace coal, oil and gas, we cannot conjure plants that are able to make useful energy available to us—as fossil fuel reserves have effectively done for centuries—on an immediate basis.[71] Instead, we must be patient as the clock of solar time is "wound up" by plants, and the combustible materials that we seem so unable to cope without are brought into being.[72] Being made to wait, moreover, may just offer scope for deeper reflection on the cadence of our relationships with energy resources, and the root causes of our desire always to "maximize production by consuming ever more power and deploying it ever more efficiently".[73]

In short, though we may not recognise it, we face a stark choice about the role that bioenergy should play in shaping the terms of future social and economic life. Either we can try to squeeze as much power from plants as possible, pinning hopes of unlimited future growth and prosperity onto the ostensibly irrepressible regenerative capacities of the vegetal world, or we can conspire with plants to imagine new kinds of relationships with resources—and resource limits—predicated not around continually expanding consumption, but rather a more deliberate embrace of post-productivist alternatives.[74] As political ecologist Giorgos Kallis has recently contended, "[C]apitalism's imaginary of a permanent transgression of a constantly expanding frontier ... is not only self-destructive; it is not fun".[75] The dormant but nonetheless inextinguishable promise of bioenergy, then, in spite of all the deleterious consequences that its modern incarnations threaten to bring about, and quite against the productivist visions of its most vocal protagonists, is that a closer attunement to plants could yet provide opportunities for new forms of energopower to emerge in post-fossil energy societies—forms of energopower dedicated not to speeding things up or even to making things more sustainable, but rather to the promotion of altogether more nuanced forms of human flourishing, satisfaction and indeed pleasure.

Notes

1 International Energy Agency, *Renewables 2023*, 94.
2 International Renewable Energy Agency, *Global Renewables Outlook*, 29.
3 Proskurina *et al.*, "Global biomass trade for energy," 375.
4 Glasgow Declaration on Sustainable Bioenergy, *Sustainable Bioenergy at the Heart of Global Net Zero*, 14.
5 Parajuli *et al.*, "Towards oilcane," 476.
6 Intergovernmental Panel on Climate Change, *Global Warming of 1.5°C*. For comparison, according to recent estimates, annual global emissions of carbon dioxide in 2018 totalled just over 41 billion tonnes (see Committee on Climate Change, *Net Zero*, 58).
7 International Renewable Energy Agency, *Global Renewables Outlook*, 80. The prefix "modern" is routinely used to distinguish contemporary, large-scale systems of bioenergy production, distribution and use from ostensibly more "traditional", smaller-scale equivalents, including by major intergovernmental organisations such as the International Energy Agency (IEA), the International Renewable Energy Agency (IRENA) and the Intergovernmental Panel on Climate Change (IPCC) (see for example Chum *et al.*, "Bioenergy"). Scholars advocating a more ethnographic approach to the study of bioenergy cultures are critical of this distinction, however (see note 11 to this chapter).
8 These comments were made by the executive director of the International Energy Agency, Dr Fatih Birol, in 2018 (see International Energy Agency, "Modern bioenergy leads the growth of all renewables to 2023").
9 Throughout the book, I deploy the adjective "vegetal" interchangeably with the adjective "plant", in line with a growing strand of scholarship in geography and critical plant studies. See for example Head *et al.*, "Vegetal politics"; and Lawrence, "Listening to plants."

10 As Kristin Asdal and colleagues put it, "[B]ioeconomies are the oldest of all economies, the economy that we have and that we also still heavily rely on." Asdal *et al.*, "'The good economy'," 19.
11 Chatti *et al.*, "Exploring the mundane," 28. Chatti and colleagues are critical of the popular use of the terms "modern" and "traditional" to categorise energy technologies and fuels, arguing instead for the concept of "mundane bioenergy" because "too often, things that are perceived to be old, unchanging, or technologically un-exciting get overlooked despite their relevance to many people" (28). The idea of a "traditional" biomass cookstove, for example, has no obvious emic equivalent and is typically imposed externally by researchers and development actors seeking not to engage with mundane energy issues on their own terms, but simply to hasten a transition towards more "modern" alternatives.
12 Lohmann, "Bioenergy, thermodynamics and inequalities."
13 International Renewable Energy Agency, *Global Renewables Outlook*, 21–29.
14 European Commission, *A Sustainable Bioeconomy for Europe*; and The White House, *Bold Goals for US Biotechnology and Biomanufacturing.*
15 Asdal *et al.*, "'The good economy'."
16 As will be discussed in more detail in Chapters 1 and 2, claims around the supposed "carbon neutrality" of bioenergy are deeply contested.
17 Buck, *After Geoengineering.*
18 Huber and McCarthy, "Beyond the subterranean energy regime?" 663.
19 Daggett, *The Birth of Energy*, 27.
20 Lenton and Latour, "Gaia 2.0," 1067.
21 Jevons, *The Coal Question*, 141. Jevons goes on to remark that "in our [coal] seams we have peculiar stores of force collected from the sunbeams for us" (141).
22 Huber and McCarthy, "Beyond the subterranean energy regime?"
23 See Simpson and Szeman, "Impasse time," and Chapter 2 for a fuller discussion.
24 Lohmann, "Bioenergy, thermodynamics and inequalities," 97.
25 Plants are not formed principally of hydrocarbon molecules themselves, of course, but rather of carbohydrates (while the former comprise only hydrogen and carbon atoms, the latter contain oxygen atoms as well). Nonetheless, the fact that carbohydrates can be readily converted into hydrocarbons—for example, through fermentation or pyrolysis—permits an energetic view of plants as a viable basis for large-scale hydrocarbon fuel production.
26 Mitchell, "Carbon democracy," 401; and Jonsson, "Abundance and scarcity in geological time," 91. Jonsson traces ideas of coal as a resource assembled in "deep time" to Victorian era debates around the possibilities of coal exhaustion in the UK. See also Sieferle, *The Subterranean Forest.*
27 For more detailed accounts of these dynamics, see Mitchell, *Carbon Democracy*; Huber, *Lifeblood*; and Malm, *Fossil Capital.*
28 Boyer describes fossil fuels as "primary instruments for the monopolization of political authority, thus constituting ... an energopolitical apparatus reinforcing both the inertia of a particular organization of fuel and a particular organization of state-based political power." Boyer, *Energopolitics*, 16.
29 Daggett, *The Birth of Energy.*

30 This is not to detract from the historical significance of plant-based industrial production systems outside of the energy sector, such as those linked to rubber or jute, for example (see Harp, *A World History of Rubber*).
31 This overarching approach to modern bioenergy is partly inspired by Jessica Dempsey's analysis, in the realm of biodiversity conservation, of the notion of "enterprising nature" as a heterogenous apparatus that animates diverse knowledge making and governance practices. Dempsey, *Enterprising Nature.*
32 Foucault, "The confession of the flesh," 194.
33 Scott Kirsch and Don Mitchell, drawing on the writing of Karl Marx, explicitly describe fossil fuels as a repository of "dead labor". This idea will be developed more fully in Chapter 1. Kirsch and Mitchell, "The nature of things."
34 On the restructuring of rural–urban relationships, see McDonagh, "Rural geography III." On shifting energy resource geopolitics, see Scholten *et al.*, "The geopolitics of renewables." On the potential for renewable or post-carbon energy agendas to perpetuate racist and colonial hierarchies, see Lennon, "Decolonizing energy" and "Postcarbon amnesia."
35 Fish, *Living Factories*; and Smith, *Uneven Development.*
36 Birch, *Neoliberal Bio-economies*; and Cooper, *Life as Surplus.*
37 O'Connor, "On the two contradictions of capitalism."
38 Prudham, *Knock on Wood*, 13.
39 McCarthy, "A socioecological fix to capitalist crisis and climate change?"; and Ekers and Prudham, "The metabolism of socioecological fixes."
40 Boyd *et al.*, "Industrial dynamics and the problem of nature."
41 Castree, "Neoliberalising nature"; and Apostolopoulou, *Nature Swapped and Nature Lost.*
42 Smith, "Nature as accumulation strategy."
43 Mitchell, *Carbon Democracy*; and Huber, *Lifeblood.*
44 Huber, *Lifeblood*, xvi.
45 Huber, *Lifeblood*, xv.
46 Boyer, "Energopolitics and the anthropology of energy"; and Boyer, *Energopolitics.*
47 Fitzsimmons, "The matter of nature"; and Castree, "Nature is dead!"
48 Kloppenburg, *First the Seed*; and Bakker, *An Uncooperative Commodity.*
49 Marx, *Capital: A Critique of Political Economy*, 1:279.
50 Braun, "Environmental issues," 669.
51 Bakker and Bridge, "Material worlds?" 19; see also Richardson and Weszkalnys, "Introduction: Resource Materialities."
52 Battistoni, "Bringing in the work of nature." On the production of broiler chickens, see Beldo, "Metabolic labor"; on the labour of lions in conservation economies, see Barua, "Nonhuman labour, encounter value, spectacular accumulation"; and on "lively commodities" in the exotic wildlife trade, see Collard and Dempsey, "Life for sale?"
53 Haraway, *When Species Meet.*
54 Barua, "Animating capital," 650.
55 Since he regards energy as integral to the productive forces of capitalism, this view of the first contradiction as fundamentally ecological is shared by Huber in *Lifeblood.*

56 On microbial labour, see Krzywoszynska, "Nonhuman labor and the making of resources"; and on vegetal economies, see Ernwein *et al.*, *The Work That Plants Do*.
57 Palmer, "Putting forests to work?"
58 Cooper and Waldby, *Clinical Labor*, 12; see also Birch and Tyfield, "Theorizing the bioeconomy."
59 Federici, *Wages against Housework*; and Besky and Blanchette, *How Nature Works*.
60 Barua, "Nonhuman labor, encounter value, spectacular accumulation," 283.
61 Moore, *Capitalism in the Web of Life*.
62 Besky and Blanchette, *How Nature Works*, 5.
63 For a discussion of the "accelerationist" tendencies of both ecomodernism and visions of a circular economy today, see Daggett, *The Birth of Energy*, 190–191; and Zwier *et al.*, "The ideal of a zero-waste humanity." For an analysis of the root causes of contemporary political convergences (at least in western Europe) between the labour and ecological modernisation movements, see Barca, "Labour and the ecological crisis."
64 Head *et al.*, "Vegetal politics."
65 Myers, "From edenic apocalypse to gardens against Eden," 137.
66 Weeks, *The Problem with Work*; see also Daggett, *The Birth of Energy*.
67 As Miriam Boyer and colleagues put it in their discussion of dominant framings of biomass in bioeconomy discourses: "Biomass is itself a homogenizing notion that obscures the biophysical materiality of nature to the extent that it erases from view particular qualities of nature in favor of comparable mass or energy units." (Boyer *et al.*, "The making of sustainability," 682).
68 Ernwein *et al.*, *The Work That Plants Do*.
69 Daggett, "Energy and domination."
70 Rabinbach, *The Human Motor*, 46; and Sedgewick, *Coffeeland*, 85.
71 In reality, of course, fossil fuel reserves offer only the *illusion* of immediate energy provision, since their formation takes place over vastly slower timescales—and in a far less efficient manner—than those applying to bioenergy resources. According to one estimate, the quantity of fossil fuels consumed in 1997 alone would originally have required biotic metabolism equivalent to more than four hundred times the planet's current net primary productivity to form; see Dukes, "Burning buried sunshine."
72 While for William Stanley Jevons, the metaphor of a "wound up" spring was used principally to emphasise the prodigious energetic potential of coal, his contemporary, the Irish physicist John Tyndall, described the formation of *living* plants as a process of "winding up", to be contrasted with animal formation as a process of "running down". See Jevons, *The Coal Question*; and Tyndall, *Heat Considered as a Mode of Motion*.
73 Underwood, *The Work of the Sun*, 173.
74 Palmer, "Rethinking plant power."
75 Kallis, *Limits*, 115.

1

Dead stocks to living flows? Bioenergy as vegetal labour

> All fossil fuels were once biomass, and the prospect of dramatically shortening the time geological forces take to convert vegetation into oil, gas and coal ... now intrigues many thoughtful persons.
>
> —Denis Hayes, *Plant Power*, 1977

Burning plants is a prehistoric practice which has fuelled the growth of all human civilisations and sometimes even contributed directly to their demise.[1] Absent plants' "wily aptitude"[2] for harnessing sunlight to synthesise water and carbon dioxide into organic compounds, all manner of earthly life could not exist. Without photosynthesis, oxygen in the atmosphere would likely be virtually absent, multicellular lifeforms may never have emerged and even the concept of energy itself could not have been invented. Plants, in short, have made the world as we know it, and it is on their shoulders—or perhaps their boughs—that human societies today have little choice but to maintain a steady footing. Despite the inescapably vegetal foundations of all human societies, however, in contemporary discourses about global energy futures, bioenergy is more often defined by its "generic capacity [to act] as an alternative to fossil fuels"[3] than in relation to plants' own characteristics. Indeed, as climate warming intensifies and the clamour to abandon fossil fuels once and for all gathers strength, it is the perceived ability of so-called "modern" forms of bioenergy to imitate patterns of resource use which are already familiar—at least in industrialised, affluent societies of the global North—that takes precedence over any potential for plants to help provoke a deeper rethink of human relations with energy and the environment. To be sure, the proven power of plants to drastically alter the composition of the earth's atmosphere makes them obvious partners in ongoing efforts to deliberately recalibrate flows of carbon dioxide entering and leaving that atmosphere today. But my argument in this book is that it is not simply the mitigation of climate change, however unprecedented this threat might appear to be,[4] which is at stake as governments and corporations across the developed world embark upon a modern "bioenergy boom".[5] Deriving increasing quantities of fuel from vegetal lifeforms should also concern us, I contend, for reasons pertaining to the underlying meaning and purpose of energy itself.

The core aim of this chapter is to lay down the theoretical foundations for these arguments. In recent years, considerable ink has been spilled—by academic researchers and environmental campaigners alike—in charting the deleterious socioecological impacts of intensive, industrialised forms of bioenergy production. While such work has generated potent critiques of modern bioenergy's prodigious ongoing expansion, however, in this book I adopt an alternative analytical vantage point, one rooted firmly in the metabolic activities of plants themselves. Weaving together insights from resource geography and anthropology, the energy humanities and especially post-humanist accounts of the nature–capital dialectic, I begin by unsettling the apparently clear-cut distinction between living bioenergy and dead fossil fuels, exposing the shared reliance of both fossil and bio-based energy upon the "work that plants do" when growing, photosynthesising and sequestering carbon dioxide.[6] Accordingly, rather than viewing plants as a theoretically inexhaustible, renewable alternative to fossil fuels, in the core of the chapter I suggest instead that bioenergy should be named explicitly as the product of "vegetal labour,"[7] and hence recognised as dependent upon nonhuman work whose orientation towards specific end goals is never "natural", but always inherently open to contestation. Finally, by honing in more closely on plants themselves—that is, by foregrounding the temporalities of their metabolism and growth, their visible appearance, and the innately unpredictable nature of their generativity—I also contend that plants could yet play a decisive role not just in restructuring prevailing energy systems, but in helping to reorient prevailing energy *cultures* as well. The political potentials of plants as energy providers should not be viewed, according to this line of thought, as restricted to bringing about a renewable energy transition or even a more stable climate future, however urgent and imperative those twin objectives might seem. Rather, the most distinctive affordances of plants as energy providers, I argue, can be found in the capacities of vegetal lifeforms, if we should choose to "listen" to them,[8] to help reshape human ideas about the fundamental purpose of energy itself, with implications for prevailing societal attitudes towards growth, waste, productivity and even pleasure.

Returning to the surface? Bioenergy and the global renewables rush

In stark contrast to industrial fossil fuel burning, the widespread use of plants as fuel far predates the modern concept of energy, at least as understood through the so-called "laws" of thermodynamics.[9] Even in states where intensive fossil fuel burning has been ubiquitous since the industrial revolution, moreover, bioenergy has never simply disappeared. Instead, plant matter—whether manifesting as "waste" or "residues" left over from wider agricultural and forestry activities, or in the form of dedicated energy crops—has retained a strong foothold as a commonplace fuel source for domestic heat, small-scale electricity generation and in some cases even transport propulsion.[10] At a global scale, meanwhile, despite persistent efforts by government agencies and development actors to disincentivise the practice on health and environmental grounds, cooking with so-called "traditional" biomass fuels—such as fuel wood, charcoal or crop residues—has remained integral to meeting the basic sustenance needs of approximately three billion people, largely concentrated into the world's poorest regions.[11] The experience of the fossil fuel era, in short, has been very far from universal. Nonetheless, however far removed many communities may be from the time-spaces of fossil fuel burning, the consequences of that burning are inescapable for all. Consequently, the rapid expansion of modern bioenergy use today is typically understood to constitute an unprecedented project in which all of humanity—whether accustomed to the decadence of fossil fuel combustion or not—has a profound stake. This is a bioenergy project concerned, in plain terms, not only with powering human activity, but also with helping to bring anthropogenic climate warming under control.

For many scholars concerned with the politics and socioecological effects of efforts to mitigate climate change, moves to replace fossil fuels with resources derived from biological matter—and especially from plants—form just one branch of a much wider process of energy transition. Indeed, the growing rush to replace fossil fuels with "renewables"—whose future supply is theoretically unlimited—has precipitated a dramatic global expansion not just of tree plantations and energy crop cultivation, but also of wind farms, solar arrays, hydropower projects and more besides. To be sure, not everyone is in a rush to leave fossil fuels behind; myriad forces of "climate delay"—whether economic, political or cultural in origin—have proven obstinate even as the reality of an already-changed climate makes itself increasingly felt.[12] Yet the conundrum of how best to coordinate technological and social innovations to accelerate a low-carbon energy transition,[13] while vital, should not be allowed to distract critical scholarly attention from the lived effects of that transition on the ground. The extent to which various forms of renewable energy deliver promised reductions in GHG emissions when compared to fossil fuel burning, for example, is not guaranteed to meet any preordained threshold, and this demands careful analysis. No less important, at least from a climate justice perspective, are in-depth, place-specific examinations of the wider social and environmental consequences of renewable energy projects.[14] After all, for all the technological complexity and diversity that renewable energy resources and their associated infrastructures embody, they are united by perhaps their most prosaic characteristic—their spatial extensiveness, and hence their requirement for land. Such an observation might appear, at first glance, so obvious as to be unworthy of deeper reflection. Yet as geographers Matt Huber and James McCarthy point out, as industrial societies move away from "a vertical reliance upon subterranean stocks" of energy and towards spatially extensive renewable alternatives, this raises "important political questions about existing land-use patterns and livelihoods, particularly in rural areas".[15] An increasingly renewable energy regime thus promises to bring political struggles and questions over who should control energy production, and for what ends, to the surface—quite literally unearthing the logics by which human relations with energy are structured, and exposing them, potentially at least, to fresh forms of critique and resistance.

Where bioenergy specifically is concerned, critical scholars and campaigners have spared little effort in highlighting the deleterious consequences of intensive energy crop cultivation regimes, whether socially, ecologically or for the climate itself. Countless detailed accounts have been assembled, especially in global South contexts, of biofuel—or, for some, "agrofuel"—production riding roughshod over pre-existing local strategies for living and working with plants.[16] Even the mere promise of future profitability attached to biofuels, acting as a magnet for financial speculation and investment, has routinely been linked to widely perceived accelerations, in such settings, of food insecurity, land enclosures and rural dispossessions—the latter often enacted on dubious legal grounds—since the turn of the century.[17] Ecologically, meanwhile, biofuel mandates and subsidies implemented from the early 2000s onwards, including prominently by the UK, EU and US, have been linked to perverse climate and biodiversity outcomes arising—among other things—from the drainage of peatlands to establish palm oil plantations in South East Asia,[18] from increased fertiliser run-off into major rivers in the American Midwest[19] and from the degradation of forests in locations as far flung as Latin America, the American South and Eastern Europe.[20]

On top of all this, the contention that using bioenergy does benefit the global climate system has itself been staunchly critiqued. To begin with here, critics point out that while trees and crops may well absorb carbon dioxide as they grow, considerable energy inputs are normally required to harvest, transport and process those plants into reliably combustible fossil fuel substitutes.[21] Where bioenergy crops are grown on lands not previously used for agriculture, meanwhile, the process of land-use change itself can also cause additional "large releases of carbon", incurring a "carbon debt" with potentially long payback times, particularly if plantations replace natural forests or grasslands.[22] Nonetheless, under international GHG accounting norms established by the United Nations Framework Convention on Climate Change (UNFCCC), biomass fuels are effectively classified as "zero carbon" at the point of combustion, with emissions linked to biomass burning instead being counted in the land-use sectors of countries where biofuels are grown.[23] This state of affairs is mirrored in prominent EU legislation on renewable energy, and has been described by some commentators as a "critical climate accounting error" akin to "counting the GHGs from coal-mining machinery but not from burning the coal".[24] Ultimately, this is an approach which discounts the carbon dioxide emissions released when biofuels are burned, even though there is no guarantee that new trees or crops will be replanted to help reabsorb similar volumes of carbon dioxide again in the future.

Existing critiques of modern bioenergy production—as socially toxic, environmentally counterproductive or both—are undoubtedly persuasive, and have been pivotal in forcing policymakers to revise the scope of biofuel mandates to limit the consumption of feedstocks which violate specific "sustainability criteria" or other supply chain safeguards.[25] Yet concerted resistance to the structural political economy of modern bioenergy production—that is, to the tendency for large swathes of land to be concentrated into the hands of a limited roster of corporate energy interests, for whom plants can then form the basis for significant wealth accumulation and profit—has proven elusive in practice. For political theorist Andreas Malm, the ongoing global transition away from fossil fuel resources and towards renewables is notable principally for the challenge it allegedly presents to incumbent economic and political power structures around energy.[26] From this perspective, fossil fuels constitute "stock" energy resources—materially concentrated and spatially fixed—which have typically been highly conducive to centralised forms of control over energy production and supply, and hence over human conduct more broadly. A growing move towards so-called "flow" energy resources formed by the sun, wind, rivers and tides opens up the possibility, by contrast, of more diffuse, distributed and hence democratic energy systems, over which the grip of large corporate actors might be significantly loosened.[27] The prodigious growth of community renewable energy schemes alone, while an admittedly crude measure of the merits of this hypothesis, cannot be ignored as evidence that flow energy resources do have radically different political affordances to coal, oil and gas. One virtually never hears of community *fossil* energy. Yet, if we are indeed living through a period of upheaval constituted by the widespread reorientation of developed economies not around static "stocks" of energy, but dynamic, distributed "flows", the place of bioenergy in this set of shifts is far from unambiguous.

Certainly it is true that the industrial cultivation of biomass for energy generation can be responsible for social injustice and ecological degradation, just as is the case for all renewable energy projects.[28] Yet the idea that bioenergy should be classified as one among many kinds of renewable or flow-based energy resources—to be compared strictly according to their ability to reduce GHG emissions associated with energy production while avoiding, at least as far as possible, wider forms of collateral damage—is contestable in two important respects. Firstly, at the landscape level, the visible hallmarks of bioenergy production are often indistinguishable from longer-standing forms of intensive arable agriculture or indeed plantation forestry. In other words, bioenergy production—at least to the untrained eye—is far from easy to spot amidst rural landscapes already deeply moulded by intensive food, fodder and timber production.[29] Perhaps as a result, even where critiques of bioenergy production do push back against the consolidation of corporate economic power, land ownership and concomitant dispossession of local communities, these processes are often categorised as specific examples of a wider, evolving political economy of intensive plantation agriculture rather than as novel political economies of energy per se.[30] By contrast, solar arrays, wind farms and hydropower projects leave onlookers in little doubt as to their expressly energetic purposes. Where local communities and concerned outsiders join forces to highlight the sometimes-dark underbellies of these renewable schemes, therefore, their critiques typically gesture more assertively towards the need for alternative, more democratic modes of organising and sharing the benefits of energy production itself.[31]

Yet if modern bioenergy's superficial resemblance to intensive agricultural and plantation forestry regimes acts to deflect attention from questions about the structural political economy of energy production specifically, this is by no means the only fissure along which bioenergy's politics diverge markedly from those of solar, wind or other flow-based renewables. The second and more immediately apparent difference, very simply, is that bioenergy is unique among renewables in deriving from the metabolic activity of living organisms. Solar radiation, atmospheric circulation and hydrological or tidal activity may all be theoretically inexhaustible energy resources, but they are all also abiotic—devoid of life. By contrast, bioenergy—whether garnered from crops or trees, or from other organic matter—inevitably relies upon the ability of living organisms themselves to make use of the primary abiotic flow energy resource—solar radiation. Even so, to fetishise the living origins of bioenergy as its key distinguishing characteristic would—I argue—be a missed opportunity. After all, at the point of actual use, bioenergy does not manifest physically as a flow of any kind, but rather as a carefully stabilised, non-living object waiting to be deployed in exactly the same way as a lump of coal, a gallon of oil or a canister of natural gas. Like all fossil fuels, bioenergy takes the form of a combustible hydrocarbon which can be easily transported, stored away for later use and accumulated into large reserves if necessary. Despite its widespread depiction as a decarbonisation tool, therefore, bioenergy does nothing to eradicate carbon from global energy systems by itself. Instead, its key contribution is to rework "the geographies and temporalities of the carbon metabolism" upon which incumbent energy systems rely.[32] Put differently, while the ability of modern bioenergy to reduce carbon dioxide emissions when compared to fossil fuels may be hotly contested, there can be no disputing the fact that today's prevailing regulatory logic of modern bioenergy aims to reinforce, rather than call into question, practices of widespread, industrial-scale hydrocarbon combustion which find their origins in the fossil fuel era.[33] Taken collectively, these ambiguities suggest that if we are to pinpoint the unique political affordances of bioenergy, we must reach beyond existing, binary theoretical distinctions, whether between fossil fuels and renewables, or indeed between static energy stocks and dynamic energy flows.

Theorising bioenergy: from vitalism to productivism

If critics of modern bioenergy production increasingly raise concerns about its uncanny likenesses—materially, infrastructurally and politically—with fossil fuel-based energy systems, protagonists rarely neglect to ground their own more optimistic appraisals in plants' status as dynamic, living organisms. The supposedly indefatigable capacities of vegetal lifeforms to regenerate, conjure their own substance "out of thin air"[34] and hence transcend the limits of resource depletion associated with fossil fuels, from this perspective, should be sufficient by themselves to secure unequivocal approval for plants' increased use as energy resources. To derive more of our energy from plants, rather than from mineral deposits buried deep below ground, appears here as a matter of basic common sense—put simply, why rely upon fuel resources which will one day run out, when better use could still be made of the abundant "free gifts" continuously being offered up by nature at the earth's surface?[35] The spectre of global climate change, while of more recent vintage than concerns about resource depletion, offers little resistance to these arguments. On the contrary, it is precisely because plants are alive—because they make use of solar energy arriving at the earth's surface in real time—that they are able, at least in theory, to preserve a delicate balance between the amount of carbon dioxide removed from and reintroduced to the atmosphere by the diverse human activities that they power. Of course, the replacement of petrochemical hydrocarbons with biologically derived ones undeniably does involve substituting dead matter with what might be described as a living fuel. Moreover, visions of a plant-powered world no longer encumbered by anxieties over resource toxicity, depletion or conflict are far from new. Holly Jean Buck, for example, discusses the failed American "chemurgy" movement of the 1930s and 1940s, for whose enthusiasts "plant resources also offered the dream of universal abundance and a peaceful world".[36] Nonetheless, even in the light of present-day concerns about climate warming, striking a balance between carbon dioxide leaving and entering the atmosphere only emerges as a concern in the first place if continued large-scale requirements for hydrocarbon fuels are taken for granted. To be sure, the fact that plants are alive matters greatly to the politics of the energy systems they can be made to sustain. Yet to presume that bioenergy's animacy alone renders it inherently desirable is—I argue—symptomatic ultimately of a deep-seated conflation between ideas about the energetic basis of life and ideas about the central purpose for which life itself is meant to be lived.

To associate energy with life is hardly a heterodox move. In charting the pre-Victorian, philosophical forebears of modern understandings of energy, for example, Cara Daggett highlights the formative influence of German polymath Gottfried Leibniz's conviction that the essence of all substance in the universe can be found not in the measurable properties of its spatial or temporal extent (as would have been argued by Descartes), but rather in the more intangible magnitude of its so-called *vis viva*, or "life force".[37] Marx, though admittedly more communist than cosmic in outlook, evokes a similar idea to Leibniz in referring to human labour power as a derivative ultimately of a person's "vital force".[38] Even the development of a modern science of life—biology—has been shown by Michel Foucault to have been predicated on a turn away from the visible, surface-level characteristics and differences between species as the predominant basis for their taxonomic classification, and towards instead "the internal laws of the organism".[39] As a result, and as Jeffrey Nealon eloquently puts it, in biology, "life is not representable ... [it] is in fact a kind of unplumbable depth, animating the organism from a hidden origin somewhere within".[40] Much like energy itself, then, life as conceived in modern biological terms cannot be seen or held in the palm of one's hand.[41] But even if life and energy share an ambiguous status in this respect—at once key building blocks for vast fields of human knowledge, yet inherently ungraspable on their own terms—this alone cannot account for widespread views of bioenergy as an inherent force for societal and environmental good. Instead, what matters more precisely is the close relationship between modern understandings of life and energy on the one hand, and economic theories of value production under capitalism on the other. To linger with Foucault a moment longer, more significant than modern biology's move to locate life at the level of internal processes, hidden within the organism, is the manner in which this shift was paralleled, in the late eighteenth and early nineteenth centuries, by developments in classical political economy serving—as Melinda Cooper puts it—to relocate "the source of all wealth behind the surface effects of exchange, in the time-processes of force, labor, and fatigue".[42] Put differently, in Foucault's "archaeology" of the ostensibly unrelated human sciences of biology and political economy, it is in both cases a turn to hidden, unrepresentable forces—of organic growth and human exertion, respectively—which promises to unlock the secrets of where life and value themselves actually come from.[43]

From the early nineteenth century onwards, then, just as life itself comes to be understood "as a process of evolution and ontogenetic development"—as a process of growth—human exertion, harnessed as labour power, comes to be understood as the principal source of value, and so as the core basis for the growth of the economy.[44] Henceforth, it is not simply life but also prosperity, wealth and human progress writ large which are recognised as originating from hidden, unrepresentable forces. Small wonder that it is also in the nineteenth century that a new Western science of those hidden forces—a science of *energy*—first emerges to help bring greater order, dignity and control to those seeking to manage and improve both life and the economy alike.[45] On the surface, of course, such developments might be viewed as nothing more than an innocuous coincidence, or even as an inevitable consequence of continuous advances in human knowledge. Yet the establishment of a shared energetic basis for understanding life and the economy has normative as well as descriptive implications. In elaborating the concept of the biosphere in the early twentieth century, for instance, Russian geochemist Vladimir Vernadsky—as succinctly captured by Melinda Cooper—defined life ultimately according to "its ability to transform solar radiations into new chemical compounds, thus accumulating a relentless surplus of 'free energy' over and above the chemical deposits already available on earth".[46] Such a definition, crucially, doesn't merely depict life as fundamentally energetic; it also imbues life with unashamedly capitalist traits—not least the capacity to accumulate a surplus effectively for free. Attributing life and economic growth alike to hidden energetic forces, it might thus be argued, serves to conflate descriptive questions about how life and the economy are made on the one hand, with normative questions about what life and the economy should be for on the other. More specifically, the nineteenth-century Western science of energy conflates vitalism with productivism, invoking the nature of life as justification for the pursuit of continuous economic expansion and growth, while making the pursuit of continuous economic expansion and growth the primary purpose of life.[47] Ultimately it is this conflation, rather than the animacy or "liveliness" of plants per se, which I argue underpins views of modern bioenergy as an inherently desirable project for humanity today.[48]

How then to counter these ideas and their influence in debates about global energy futures in a supposedly post-fossil fuel era? To break the hold of such thinking requires in the first instance a reinterrogation of the neat "discursive line between clean, green, lively bioenergy and dirty, black, dead fossil fuels".[49] Viewed from the vantage point of the industrial revolution and its aftermath, such a distinction appears rock solid. From a significantly extended temporal perspective, however, the seemingly impermeable boundary between "organic" and "minerals-based" energy resources begins to break down.[50] After all, fossil fuels are but the geological remains of formerly living plants and the animals which fed upon them—in other words, the long-buried relics of prehistoric photosynthesis and carbon dioxide sequestration. These inextricably photosynthetic origins of fossil fuels are far from a new discovery, of course. Frederick Albritton Jonsson's analysis of nineteenth-century debates over the possibility of coal exhaustion in the UK, for instance, reveals prevalent Victorian understandings of coal as a kind of "providential gift, produced by a forested world now irrevocably lost", and hence as a basis for bringing humanity "into immediate connection with the vegetation that clothed the ancient earth".[51] More recently, such ideas have found echoes in pithy descriptions of coal seams as a kind of "subterranean forest", or fossil fuels more broadly as reserves of "buried sunshine".[52] Though apparently playful, the implications of these observations for efforts to distinguish bioenergy from fossil fuels are profound. Indeed, a "deep time" perspective belies the perceived importance of the liveliness and productivity of plants when compared to the inert, dead nature of coal, oil and gas. What separates these resources in the end is not vitality, but time—the key difference lying not between life and death, but between photosynthesis enacted by plants and phytoplankton consuming solar energy delivered to the earth hundreds of millions of years ago, and photosynthesis enacted by plants making use of "immediate circulations of solar energy" today.[53] To describe global energy systems today as being fuelled increasingly by plants, from this longer-term perspective, therefore makes little sense. Powering human social and economic activities with plants is virtually all we ever have done.

Perhaps the most obvious rebuttal to attempts to flatten the conceptual differences between bioenergy and fossil fuels can be found in the view that plants differ from coal, oil and gas not merely by being alive, but by being amenable to deliberate modification and improvement. The charge of a naive vitalism, from this perspective, can be dismissed as a misguided caricature of a sober, science-based commitment to actively improve upon nature by transforming it. Certainly, dominant policy visions and scientific assessments of the scale of the global energetic resource that plants might embody in the twenty-first century are ambitious not just because new plants can normally be relied upon to replace old ones. The innate capacity of plants to regenerate is only half the story; what appeals just as much, if not more, is the potential for new plants also to be made—through deliberate interventions operating at scales ranging from the molecular to the landscape level—to grow more vigorously and robustly than their predecessors. That plants' living status offers a "distinct avenue for productivity advance through 'improving' nature directly"[54] is already evidenced extensively, not least by the considerable success of diverse "yield improvement" programs which have long formed a staple component of profit maximisation in industrial agriculture and forestry alike.[55] Yet such programs must also respect what Melinda Cooper describes as the "one limit that capitalism never escapes"—the imperative to derive profit.[56] Consequently, even as yield improvements have served, at least in some contexts, to partially remediate concerns about resource depletion, those same programs have also driven the imposition of new kinds of limits altogether. These new limits manifest, for example, in the restricted allocation of intellectual property rights over those forms of scientific knowledge and technological innovation from which yield improvements are derived. They can also be observed in the changing natures of vegetal lifeforms themselves, as highlighted by Jack Kloppenburg's assessment of hybrid crop cultivars as being expedient to capital not just for their propensity to drastically increase yields, but also for their sterility, which prevents farmers from harvesting additional viable seeds from their own produce effectively "for free".[57] While capitalist production based upon biological processes is undoubtedly committed to improving and optimising nature, therefore, it also presents a contradiction—to safeguard the generation of profit, this production "needs to depotentialize the future possibilities of life, even as it puts them to work".[58]

Moving to cultivate more and more biomass expressly for the purposes of energy generation—and not simply to produce food, timber or other bio-based commodities—does nothing by itself to eradicate the fundamental tension, inherent to all "bioproduction", between the promise of a drastically augmented resource base on the one hand, and the need to restrict resource ownership and availability as a means of extracting profit on the other. Indeed, when compared to fossil fuels, modern bioenergy might be argued to appeal no more for its alleged propensity to eradicate societal concerns about resource depletion, than for the theoretical ability of plants—as living resources—to furnish capital with greater and more precise degrees of control over the *extent* of resource scarcity than those which have historically been afforded by dead reserves of coal, oil or gas. Even this interpretation, however, overlooks the ways in which bioenergy may also serve to perpetuate other kinds of control, less over the biophysical extent of global energy resources than over prevailing ideas about what those resources should ultimately be for. In appearing to physically embody processes of continuous production and growth, plants may be easily upheld as paragons of a broader view that positions production and growth themselves as the defining hallmarks of a life well lived. Viewed from this perspective, modern bioenergy projects are not only concerned with pressing plants into service as a reliable basis for extracting profit from energy provision. More fundamentally, they may also be viewed as a means of reinforcing prevailing views of human progress as inextricably dependent upon the ever more efficient and productive consumption of larger and larger amounts of energy.[59] In other words, even as the industrial-scale enrolment of plants as modern bioenergy providers promises to hasten the demise of an era of fossil fuels, it also acts—at least in the vast majority of incarnations observable today—to naturalise the economics, politics and cultures of resource use which have shaped the era of fossil *energy*. Understood as an apparatus—as a "system of relations" holding diverse regulatory decisions, institutions, discourses, scientific research agendas and political programs together—modern bioenergy looks to vegetal lifeforms in the end to take on a whole repertoire of functions previously performed by fossil fuels, operating on two distinct levels: firstly as physical energy carriers, and secondly as metaphorical vehicles for specific ideas about the innate virtuousness of increasingly efficient and productive energy use. But even if modern bioenergy logics are at their core not so much vitalist as productivist, it remains rare for concerted scholarly attention to be paid to the differences between vegetal life and vegetal work.

Putting plants to work: bioenergy as vegetal labour

Critical accounts of the relationship between nature and capitalist production frequently uphold a fundamental distinction between biologically and nonbiologically based industries. According to this view, nonbiological materials are inherently resistant to change and can only be subsumed to the needs of capital in limited, "formal" terms, as passive (and not seldom obstinate) resource inputs. Biological systems, by contrast, hold out the tantalising potential to be reconfigured and further optimised for the purposes of capitalist production.[60] As a result, they enable producers to achieve what William Boyd and colleagues famously termed the "real subsumption of nature"—where nature is effectively "(re)made to work harder, faster, and better" for capital.[61] Especially where theories of resource governance are concerned, the concept of "real subsumption" has enabled biological nature to be recognised as presenting a range of opportunities and surprises—and not merely obstacles—to capitalist value production.[62] In so doing, it has arguably presaged more recent, so-called "new materialist" work, wherein the "generative capacities of biophysical processes" and the inherent "inventiveness" of life are recognised as constitutive components of resource-making itself.[63] For those seeking to move beyond a strict dichotomy between nature on the one hand and society or capital on the other, new materialism has appealed especially for its avowedly posthuman understanding of agency as a relational achievement linked more to "the coming together of things" than to the capacity of individual subjects to act.[64] Yet an equally fertile route around the nature–society dichotomy might be found, I contend, in the heritage of the real subsumption of nature framework itself, as a scion ultimately of Marx's own distinction between the formal and real subsumption of *labour*.[65] After all, if plants are today increasingly being made to do the work previously performed by fossil fuels, then exposing the full political stakes of these developments demands more than simply highlighting their dependence upon the innate generative capacities or inventiveness of vegetal lifeforms. Recognising nature as a constitutive component of economic processes, and not merely as a passive input to them, is principally a descriptive rather than a normative move—a means of rethinking the agencies involved in production which often leaves the end goals of production themselves unquestioned, if they are examined at all. By contrast, the contention that plant photosynthesis, growth and carbon sequestration can be understood as forms of labour power in their own right—in the case of modern bioenergy, that plants might be recognised and named as vegetal labourers contributing directly to value production[66]—offers a starting point both for rethinking the metabolic contours and rhythms of plant-based energy economies, and for scrutinising their fundamental desirability.

Describing plants as sources of vegetal labour may seem illogical or even perturbing in the light of conventional understandings of the capacity to labour as a distinctively human trait, normally viewed as inextricable from cognitive faculties of preconception and deliberate intention.[67] As Marx himself acknowledged, "we presuppose labour in a form in which it is an exclusively human characteristic".[68] Yet critical Marxist analyses of the relationships between nature and society have long kept on trial the idea not just of labour as a defining capacity of human beings, but also of labour as a process which must necessarily be linked to individual subjects of any kind. As Morgan Robertson and Joel Wainwright have observed, for example, Marx's writings emphasised that labour power "is itself a metabolic relation between nature and society: not a thing but a socionatural relation".[69] For Neil Smith, meanwhile, even if labour can be defined succinctly as a process "by which the form of nature is altered", this transformative, metabolic capacity belongs no more to humans than it does to nature writ large—to quote Marx again, in labouring, humans "can only proceed as nature does herself, i.e. [they] can only change the form of the materials".[70] To be sure, a focus strictly on metabolic processes is far from sufficient to grasp the diverse forms of work which animate contemporary economies today, predicated as they are on emotional and other "immaterial" forms of labour as much as material commodity production.[71] Accepting that the outputs of labour may not always be tangible, however, need not automatically rule nonhumans out of the prospective labour pool. At least when it comes to animals, in fact, influential contributions have already highlighted how the production of use values in sectors like farming, forestry and manufacturing has historically depended critically upon the brute force and metabolic exertion of horses, cattle or other domesticated species.[72] Other interventions, still more provocatively, have even posited that intangible, affective encounters with nonhumans—whether as domestic or exotic pets, as the living linchpins of wildlife- and conservation-based tourism or simply as visual icons used to generate demand for diverse commodities and services—all constitute sites in which surplus economic value emerges from "animal work" as well as human labour.[73]

If contemporary efforts to parse the full range of "living potentials that capital parasitizes upon in its quest to expand and reproduce" have been quick to invite a view of animals as workers, however, the idea that plants might labour remains much less widely accepted.[74] Certainly within some subdisciplines, including notably social and cultural geography, plants have previously been acknowledged as active participants in processes as diverse as urban greening, viticulture and even nation building.[75] Despite being underpinned by close empirical studies of working relationships involving both humans and plants, however, such analyses have only rarely accorded plants the capacity to labour as such.[76] To some extent this is hardly surprising. If the idea of animals as workers stretches conventional definitions of labour as the outcome of preconception and intentionality to near breaking point, then plants—the vast majority of which are sessile, apparently incapable of closing themselves off from external stimuli and frequently difficult to distinguish as individuals at all—would seem to shatter such notions altogether. Viewed from another perspective, however, the apparent incompatibility of plants with mainstream understandings of labour might just as easily be interpreted as evidence of conceptual intransigence than as proof of any underlying truth about the essence of labour itself. Indeed, for some thinkers, the principal value of plants is to be found precisely in their ability to expose the predominantly animalistic moorings of Western philosophy as themselves parochial.[77] When it comes to ideas about labour power, therefore, even if plants may exceed our attempts to discern individual from collective work, or to disentangle subconscious instinct from deliberate intention, this does nothing to undermine a view of all labour as fundamentally relational—in short, not as a capacity linked to individual subjects, but as "an activity enmeshed in heterogenous entanglements".[78] In her examination of the enrolment of plants within new regimes of urban park maintenance in Switzerland, for example, Marion Ernwein avers that "laboring relations" are structured through the integrated management of more-than-human "work collectives", rather than through separately delineated regimes of governance applying to human park employees on the one hand, and plants on the other.[79] By a similar token, and as will be argued in Chapter 3, efforts to maximise the production of value in the so-called "working forests" of the US South today rely no more on enhancing the interactions between individual trees as they grow, than they do on optimising interactions between trees and human workers employed to intervene in those forests at key moments in the plantation life cycle.[80]

If labour is to be reconceived, though, not as an exclusively human trait but more generously to include any "material process of transformation oriented towards the generation of capitalist value", for what purposes is it ultimately useful to acknowledge diverse living entities—whether animals, plants, algae or even microbes—as active contributors to the labour process?[81] A view of capital as dependent always upon what Alyssa Battistoni terms "hybrid labor"—the "collective, distributed undertaking of humans and nonhumans"—is not only helpful for its emphasis on interconnection and relationality.[82] Recognising that capitalism depends upon a "web of life," that it is at bottom not strictly an economic or even a social system but a "way of organizing nature",[83] is not so much the peroration of the argument I want to make here as its point of departure. Instead, what is at stake in the claim of hybrid labour, beyond the nature–society dichotomy or even the purity of species-based categorisations of the living world, is the contingency and contestability of the historically specific arrangements within which diverse forms of life are apprehended, brought to heel and enrolled within processes of economic value production. Labour, after all, is never an end in itself, but a means to an end—"a partial descriptor of lives, relationships, and activities", rather than a justification for their existence.[84] Indeed, for anthropologist David Graeber, the "real point" of Marx's labour theory of value is not strictly scientific or empirical (that is, concerned with the process by which commodity prices are determined), but "philosophical"—intended to underline that the world we inhabit is not just "something we made", but something "we could also have made differently".[85] Crucially, this possibility of difference, of working expressly to produce another kind of world, is one expunged from many contemporary and highly influential descriptions of the nonhuman realm—both within and beyond academic debates about a deepening environmental crisis—as either a repository of "natural capital" or a provider of "ecosystem services".[86] Within these instrumentalist discourses, nonhuman nature is conceived a priori as inherently subservient to existing modes of economic organisation, while the driving logic of capital itself, a logic of efficiency, productivity, expansion and "growth for the sake of growth",[87] is left entirely unchallenged. The act of recognising and labelling nonhumans explicitly as contributors to the labour process, by contrast, goes beyond emphasising the inextricably relational underpinnings of capitalist value production to question the desirability of the end goals for which work of all kinds is performed in the first place.

When it comes to plants specifically, the concept of vegetal labour therefore promises to denaturalise the diverse projects within which photosynthesis and carbon sequestration are today enrolled to help generate economic value, whether to produce renewable alternatives to fossil fuels, or indeed to advance wider "nature-based solutions" to climate change or other environmental challenges.[88] With respect to modern bioenergy in particular, insisting on a view of plants as sources of vegetal labour power serves to decentre customary concerns with the socioecological impacts of specific bioenergy projects on the ground—as important as these are—foregrounding instead the key political claim which undergirds modern bioenergy as a distinct regulatory logic. This is the claim that the deleterious environmental consequences of longstanding historical efforts to put fossil energy to work (in short, to mobilise the "dead labor" of prehistoric plants[89]) can best be countered not by pausing to consider which forms of work will actually be necessary (let alone desirable) in the future, but simply by putting more of the living world to work in place of the dead.[90] As enticing as such a claim might appear on the surface, of course, it cannot undo the fact that no manifestation of plant metabolism—whether enacted in the deep past or the present day—has ever been preordained to provide human societies with energy. Like all forms of labour, vegetal labour is never an end in itself; instead, it is a means to diverse possible ends. To conceive of bioenergy as vegetal labour thus enables critiques predicated upon evidence of negative social or ecological outcomes arising from specific instances of bioenergy production to be augmented, I argue, by a more fundamental objection to the naive vitalism at the heart of modern bioenergy as a regulatory logic. This is an objection, in short, to the idea that just because plants can be used to generate energy, their enrolment as energy resources should automatically be regarded as a virtuous act. Far from being a merely philosophical point, moreover, paying closer attention to the ways in which plants are unwittingly enrolled as co-labourers within modern bioenergy regimes entails an explicit acknowledgement both of the sovereignty of vegetal lifeforms to enact their own distinct modes of being, and of their concomitant capacity to affect us in ways unforeseen and unexpected. To speak of modern bioenergy as vegetal labour is thus ultimately not only to question the necessity or desirability of using plants to generate energy, but also to concede the power of plants to make us think again about what energy itself might really be for.

Rethinking plant power: towards alternative energy cultures?

What might it mean, then, for bioenergy to be reappraised in a manner which situates plants themselves at the centre, rather than the periphery, of our thinking? To embrace plants, in other words, not simply as an alternative, allegedly more sustainable basis for producing combustible hydrocarbon fuel than coal, oil or gas, but as potential collaborators in an effort to more fully redefine the roles that energy can and should occupy within human social and economic life in the twenty-first century? This book by no means claims to chart an exhaustive set of ways in which contemporary energy cultures might be reshaped by a closer attunement and openness to the nature, capacities and potentials of plants. Nonetheless, it does suggest that thinking more carefully about and with vegetal lifeforms could stand to productively unsettle several sets of ideas about energy which have become taken for granted in Western industrial societies, pertaining among other things to questions of scarcity and waste, availability and immediacy, and productivity and growth.

As Jonsson's examination of nineteenth-century debates about coal exhaustion in Britain bears witness,[91] the finitude of fossil fuels began stoking fears about the long-term feasibility of modern industrial societies almost from the dawn of industrialisation itself. Today, dominant policy imaginaries of the global energy system remain heavily influenced by such fears, evidenced by the prevalence of concepts like "peak oil", "energy security" and even "keeping the lights on".[92] The steady creep of fossil fuel resource depletion, however, does more than simply raise the spectre of a significant energy supply crisis, always lurking—or so it seems—just over the horizon. At a more mundane level, societal attitudes to energy use also owe much to the fact that fossil fuels can only be burned once, after which they are then gone (at least on timescales meaningful to humans) for good. The permanently lost potential for fossil fuels to be used for other ends—or, in more formal economic terms, their "opportunity costs"—have long normalised a ceaseless and frenzied quest to squeeze as much as possible out of every drop of oil or gram of coal. Indeed, as Cara Daggett has shown, the opportunity costs of fossil fuel burning have exerted a profound influence over ideas about productive resource use dating back to the Victorian era, elevating the twin goals of maximising work and minimising waste to a near-sacred status.[93]

Bioenergy cannot fully transcend these concerns about resource depletion. Whichever way you look at it, organic matter—almost always derived from plants—must be destroyed if bioenergy is to be consumed. Yet it is also true that new crops and trees can, given certain provisions, be regrown to supply fresh organic material from which further bioenergy can be derived, usually on timescales much shorter than the average human lifespan. Plants thus offer the prospect of a world in which the combustion of hydrocarbon fuels no longer strictly needs to be freighted with any regret about permanently forgoing alternative possibilities of energy use. Indeed, when set against fossil fuels, the regenerative capacities of the vegetal kingdom might even be said to actively forgive profligacy and frivolity, posing a direct challenge to prevailing ideas about what sorts of energy use should be conceived of as efficient, productive or indeed wasteful in the first place. Within real-world energy policy discourse, of course, such ideas are presently all but unthinkable. As we have seen, so detached are prevailing regulatory frameworks governing bioenergy use from the realities of the vegetal kingdom, in fact, that biomass fuels are often effectively legally classified, despite being no less emissions-intensive than coal, oil or gas, as zero carbon at the point of combustion.[94] The particular circumstances under which carbon dioxide is actually drawn down from the atmosphere by plants used to produce biomass—whether involving loblolly pine forests in the US South, sugarcane plantations in Brazil or wheat fields in eastern Europe—are here concealed from view in order to uphold a technically incorrect understanding of biofuels as near-identical replacements for fossil fuels, but without the emissions. Any remaining prospect of distinguishing bioenergy from fossil energy, moreover, is all but extinguished by a relentless concern with security of supply, through which the physical properties of biomass fuels—as tangible energy carriers capable of being amassed into large reserves and deployed on demand—come to be viewed as innately superior to those of supposedly more intermittent renewables like solar, wind or hydropower.

In a world that is growing steadily warmer, windier and wetter, there is however a certain irony to depicting bioenergy—dependent as it is upon the continued productivity of increasingly stressed global agricultural and forestry systems—as inherently more reliable than solar, wind or hydropower. Refocusing attention away from spaces of energy demand and towards spaces of energy availability serves to expose this idea, despite its hold over energy policy discourse, as absurd. "Flow-based" renewables, to again borrow Andreas Malm's terminology, are only intermittent to the extent that the weather in any given place is changeable;[95] on a planetary scale, solar radiation, wind and hydropower are available to such a continuous, widespread (and in some cases also a growing) extent that all existing human demands for energy pale into insignificance. Bioenergy, by contrast, is never truly immediately available. Rather, crops and trees stand apart from other renewable resources in acting—for the bulk of their lives, at least—as what Augustine Sedgwick has termed "quiet monuments to nature's power to make men wait".[96] Regardless of how much their vigour and indeed resilience to environmental stress may be deliberately optimised and enhanced, plants still take time to grow, and thus demand our patience as we wait for solar energy, water and carbon dioxide to undergo a process of "elemental rearrangement" yielding up combustible organic matter.[97] In short, if the capacity to forgive supposedly frivolous or profligate uses of energy is one theoretically shared by all renewable energy resources, where bioenergy is concerned this forgiveness can only ever be meted out slowly, rather than instantaneously. Plants are thus unique in offering us a source of energy which is at once neither finite nor immediate.

Being made to wait for the clock of solar time to be "wound up" by plants, however, might just offer scope for deeper reflection on the cadence of our relationships not just with bioenergy resources, but energy of all kinds.[98] After all, even if plants might seem on the surface to act as natural embodiments of economic ideals of continuous expansion and growth, in reality they are not always looking to work harder, faster or better. Over the course of a tree's lifetime, for instance, bouts of rapid, prodigious growth—especially in youth—quickly give way to much longer periods of slow, only gradual expansion, or indeed considerable periods of effective stasis (and this is to say nothing of seasonal metabolic rhythms which see the vast majority of annual and indeed perennial plants pass through distinct stages of dormancy and activity each year). Perhaps most fundamentally, however, it is worth noting that plants do not strictly produce energy at all—rather, they bring it to rest. While the leaves of plants and trees may sometimes be referred to as nature's solar panels, there is a world of difference between photovoltaics, which convert solar energy directly into electricity, and photosynthesis, which uses solar energy, carbon dioxide and water to bring stable organic matter into being, seemingly out of "thin air", as Myers puts it.[99] If we are to really learn something about energy from plants, therefore, perhaps it is that dynamism, growth and indeed ever-increasing rates of energy expenditure are not always the answer to what it means to live our lives well.

Ultimately, then, the promise of properly paying attention to plants as energy resources is that it may yet serve to expose just how unenjoyable the ceaseless pursuit of economic growth has become. Indeed, new cultures of energy use closely attuned to the nature of vegetal life, far from entailing mere asceticism and restraint, might yet give us license to creatively explore ways of working with plants to cultivate more diverse and open-ended relationships with energy writ large. This is an opportunity, in other words, to embrace plants as co-labourers in projects concerned not with fuelling permanent, global-scale commitments to productivity, growth or even necessarily sustainability, but rather with bringing into being an array of more locally self-determined modes of energy use oriented towards human flourishing, satisfaction and pleasure. These ideas will be returned to and elaborated in more detail in the book's conclusion. For now, the next chapters turn their attention to in-depth case studies of the regulatory logic of modern bioenergy as it has played out in diverse contexts over the past two decades, beginning with an account of the new rush for liquid transport biofuels which was kickstarted by American and European policy initiatives in the early 2000s.

Notes

1 See for instance Williams, "Dark ages and dark areas"; and Ponting, *A New Green History of the World*.
2 Myers, "Photosynthetic mattering," 125.
3 Chatti *et al.*, "Exploring the mundane," 33.
4 For the indigenous and environmental justice scholar Kyle Whyte, depictions of climate change as an unprecedented threat to humanity are in fact mobilised to disguise the inconvenient truth about this crisis—that it is simply "an intensification of environmental change imposed on Indigenous people by colonialism" (Whyte, "Indigenous climate change studies," 153).
5 Lohmann, "Bioenergy, thermodynamics and inequalities."
6 Ernwein *et al.*, *The Work That Plants Do*.
7 Palmer, "Putting forests to work?"; and Ernwein, "From undead commodities to lively labourers."
8 Lawrence, "Listening to plants"; and Marder, "To hear plants speak."
9 See Daggett, *The Birth of Energy*.
10 Bioethanol's widespread use as a road transport fuel in Brazil, driven by fears over energy security in the wake of oil price shocks in the early 1970s, serves as the most obvious example here. See Chapter 2 for a fuller discussion.
11 Chatti *et al.*, "Exploring the mundane," 33. The Clean Cooking Alliance, a not-for-profit organisation formed in 2010, frames the replacement of biomass cookstoves with modern cooking stoves and fuels, such as liquified petroleum gas (LPG), as a "cross cutting solution" which "directly contributes to reaching 10 out of the 17 SDGs [Sustainable Development Goals]." Clean Cooking Alliance, *The Value of Clean Cooking*.
12 Lamb *et al.*, "Discourses of climate delay"; and Carton, "'Fixing' climate change by mortgaging the future."
13 See for instance Geels *et al.*, "Sociotechnical transitions for deep decarbonisation."

14 Levenda *et al.*, "Renewable energy for whom?"
15 Huber and McCarthy, "Beyond the subterranean energy regime?" 662, 655.
16 See Bridge *et al.*, "Geographies of energy transition"; Vermeulen and Cotula, "Over the heads of local people"; and Hunsberger and Ponte, "'Sustainable' biofuels in the global South." For McMichael, the term "agrofuel" highlights "the agricultural dimensions of what are conventionally termed 'biofuels', as if such fuels represented life, as opposed to competition for scarce crop land, deforestation, and so on." McMichael, "A food regime analysis of the 'world food crisis'," 283.
17 See for example Baka, "Making space for energy"; Li, "To make live or let die?"; and Nally, "Governing precarious lives." Edelman *et al.* highlight a "research rush on the global land rush" beginning especially from 2007. Edelman *et al.*, "Global land grabs," 1528.
18 Koh, "Remotely sensed evidence of tropical peatland conversion to oil palm."
19 Costello *et al.*, "Impact of biofuel crop production on the formation of hypoxia in the Gulf of Mexico."
20 See for example Sawyer, "Climate change, biofuels and eco-social impacts in the Brazilian Amazon and Cerrado"; Schlesinger, "Are wood pellets a green fuel?"; and van der Wal, *Wood Pellet Damage*.
21 Escobar *et al.*, "Biofuels."
22 Searchinger *et al.*, "Fixing a critical climate accounting error," 527; see also Fargione *et al.*, "Land clearing and the biofuel carbon debt."
23 Chatham House, *Woody Biomass for Power and Heat*, 5.
24 Searchinger *et al.*, "Fixing a critical climate accounting error"; and Searchinger *et al.*, "Europe's renewable energy directive poised to harm global forests," 3.
25 Such schemes have attracted their own critiques, not least that framing environmental and social problems associated with globalised production networks in "terms of technical criteria and procedures might result in rendering some issues invisible and closing potential debate and contestation" (Silva-Castañeda and Trussart, "Sustainability standards and certification," 490). See also Ponte, "'Roundtabling' sustainability"; and Winickoff and Mondou, "The problem of epistemic jurisdiction in global governance."
26 Malm, *Fossil Capital*. See also Mitchell, *Carbon Democracy*.
27 Huber, himself drawing from Malm, suggests that "finding ways to profit and control wealth accumulation from such flow resources" constitutes a "real challenge for capitalism" (Huber, "Energized antagonisms," 242).
28 In-depth ethnographic research has charted such effects as they relate, for example, to wind power projects (see Dunlap, "'Murderous energy' in Oaxaca, Mexico") and solar energy (see Brock *et al.*, "Volatile photovoltaics"; and Yenneti *et al.*, "Spatial justice and the land politics of renewables").
29 As will be discussed further in Chapter 2, bioenergy's ability to blend in with existing forms of industrial production—visually as well as infrastructurally—constitutes one of its key political affordances.
30 Alonso-Fradejas, for example, decries the "corporate appropriation for free of productive and reproductive labor" in Guatemalan biofuel plantations,

yet attributes these and other unsavoury developments ultimately to "agro-extractivism" rather than to bioenergy logics specifically. Alonso-Fradejas, "'Leaving no one unscathed' in sustainability transitions," 133.

31 See for instance Dunlap, *Renewing Destruction*; Lennon, "Decolonizing energy"; and van Veelen, "Negotiating energy democracy in practice."

32 Palmer and Carton, "Carbon removal as carbon revival?" 3.

33 Lohmann, "Bioenergy, thermodynamics and inequalities."

34 While Myers writes of vegetal lifeforms "pulling matter out of thin air", this image overlooks plants' need for light, water and chemical nutrients—and not merely carbon dioxide—to be able to grow. Myers, "Photosynthetic mattering," 125. See also Lenton and Latour, "Gaia 2.0."

35 The idea that nature constitutes a "free gift" for man is deployed here in what Jason Moore calls its "emphatically critical", Marxist sense—to highlight a contradiction in capitalist approaches to the environment—even as the same notion has formed a core building block of both classical political economy and subsequent neoclassical economic theory. Moore, "Transcending the metabolic rift," 12. See also Battistoni, *Free Gifts*; and Foster and Clark, "The paradox of wealth."

36 Buck, *After Geoengineering*, 55. For Denis Hayes, writing in the late 1970s, energy crops and other organic sources of fuel constituted a promising alternative to counteract "the hastening drift into the nuclear age" (Hayes, "Plant power," 349).

37 Daggett, *The Birth of Energy*, 23–24.

38 Marx, *Capital: A Critique of Political Economy*, 1:341.

39 Foucault, *The Order of Things*, 158.

40 Nealon, *Plant Theory*, 7.

41 Indeed, if we are to hear Foucault out in full, then we must acknowledge that life—and indeed energy—are not simply reconceived at the hands, respectively, of modern biology and modern physics; rather, they are *invented* by them. As Foucault puts it: "Up to the end of the eighteenth century, in fact, life does not exist: only living beings" (Foucault, *The Order of Things*, 175). See also Nealon, *Plant Theory*, 5.

42 Cooper, *Life as Surplus*, 6. As Cooper puts it, for Foucault, "the development of modern life sciences and classical political economy should be understood as mutually constitutive events" (Cooper, *Life as Surplus*, 5).

43 One might even say that Marx's notion of the "hidden abode of production"—the arena within which human labour power is exploited, but to which entry cannot be secured "except on business"—finds a biological equivalent in the idea of life as a process located ultimately in hidden recesses of the organism. Marx, *Capital: A Critique of Political Economy*, 1:279–280.

44 Cooper, *Life as Surplus*, 7.

45 See Daggett, *The Birth of Energy*. Daggett upholds the ideas of mathematical biologist Alfred Lotka, writing in the early part of the twentieth century, as exemplary of this view of progress achieved through improved energetic governance. For Lotka, "[E]volution favoured species that increased the 'total

energy flux' of an organic system", while all energy exchanges not exploited by humanity could be "considered waste, including waterfalls and each ray of sunlight" (Daggett, *The Birth of Energy*, 119–120).

46 Cooper, *Life as Surplus*, 35.

47 For Cara Daggett, the modern Western concept of energy is significant in that it "combines a materialism, in the description of activity, with moralism, expressing a bias towards dynamism over stasis" (Daggett, *The Birth of Energy*, 18).

48 As Cooper puts it, "[T]his is a vitalism which comes dangerously close to equating the evolution of life with that of capital" (Cooper, *Life as Surplus*, 42).

49 Chatti *et al.*, "Exploring the mundane," 33.

50 Wrigley, *Continuity, Chance and Change*.

51 Jonsson, "Abundance and scarcity in geological time," 84, 89.

52 Sieferle, *The Subterranean Forest*; and Mitchell, *Carbon Democracy*.

53 Daggett, *The Birth of Energy*, 27. As Daggett puts it, "[B]urning fossil fuels is somewhat like sparking the spontaneous decomposition of a large number of long-dead plants and animals all at once" (33).

54 Boyd *et al.*, "Industrial dynamics and the problem of nature," 565.

55 See for instance Prudham, "Taming trees"; and Kloppenburg, *First the Seed*.

56 Cooper, *Life as Surplus*, 25.

57 Kloppenburg, *First the Seed*.

58 Cooper, *Life as Surplus*, 25; see also Ernwein and Palmer, "Making the mos(s)t of nature?"

59 Daggett describes such a notion as a "fossil myth of fuel expansion", according to which "humans as a species desire ever more energy, and a series of technological innovations have provided it" (Daggett, "Energy and domination," 645).

60 Boyd *et al.*, "Industrial dynamics and the problem of nature."

61 Boyd *et al.*, "Industrial dynamics and the problem of nature," 564.

62 Much work in agrarian political economy, for example, has tended to emphasise "natural obstacles" to capitalist development and industrialisation in agriculture, rather than seeing nature as a more active contributor to production itself. See for example Goodman, Sorj, and Wilkinson, *From Farming to Biotechnology*.

63 Bakker and Bridge, "Material worlds?" 19; and Braun, "Environmental issues," 667. On relational approaches to resource-making, see Richardson and Weszkalnys, "Introduction: Resource materialities"; and Kama, "Resource-making controversies."

64 Braun, "Environmental issues," 671.

65 Marx, *Capital: A Critique of Political Economy*.

66 Palmer, "Putting forests to work?"

67 The near-hackneyed juxtaposition offered by Marx is that of the "worst architect" and the "best of bees" (Marx, *Capital: A Critique of Political Economy*, 1:284). For further discussion, see Kallis and Swyngedouw, "Do bees produce value?"; and Barua, "Nonhuman labour, encounter value, spectacular accumulation."

68 Marx, *Capital: A Critique of Political Economy*, 1:283–284.
69 Robertson and Wainwright, “The value of nature to the state,” 895.
70 Smith, *Uneven Development*, 53; and Marx, *Capital: A Critique of Political Economy*, 1:133.
71 See for instance Gill and Pratt, “In the social factory?”; and Hochschild, *The Managed Heart*.
72 See for example Hribal, “ ‘Animals are part of the working class’ ”; Porcher and Schmitt, “Dairy cows”; and Beldo, “Metabolic labor.”
73 On pets, see Haraway, *When Species Meet*; and Collard, *Animal Traffic*. On encounter value in conservation tourism, see Barua, “Nonhuman labour, encounter value, spectacular accumulation.” On animals as “spectacular icons”, see Barua, “Affective economies, pandas, and the atmospheric politics of lively capital.” Work examining diverse forms of animal work across multiple spheres of human–animal interaction can be found in Porcher and Estebanez, *Animal Labor*.
74 Barua, “Animating capital,” 656.
75 On urban greening, see Ernwein, “From undead commodities to lively labourers.” On viticulture, see Brice, “Attending to grape vines.” On nation building, see Ginn, “Extension, subversion, containment”; and Ginn and Mustafa, “Planting soft Pakistan.”
76 See Head *et al.*, “Vegetal politics.” For exceptions, see Perkins, “Ecologies of actor-networks and (non)social labor within the urban political economies of nature”; Ernwein, *Les Natures de la Ville Néolibérale*; and Palmer, “Putting forests to work?”
77 Michael Marder and Jeffrey Nealon, for example, both lament a historical tendency for plants to be banished to the fringes of Western philosophy, including in most interpretations of Foucault’s concept of biopower (Marder, *Plant-Thinking*; and Nealon, *Plant Theory*). For Emanuele Coccia, meanwhile, because plants form “the most intense, radical, and paradigmatic form of being in the world ... [t]o interrogate plants means to understand what it means to be in the world” (Coccia, *The Life of Plants*, 5).
78 Barua, “Nonhuman labour, encounter value, spectacular accumulation,” 283.
79 Ernwein, “Bringing urban parks to life,” 560.
80 See also Palmer, “Putting forests to work?”
81 Krzywoszynska, “Nonhuman labor and the making of resources,” 231. On plants and algae as energy feedstocks, see Palmer, “Putting forests to work?”; and Kasdogan, *Potentiating Algae, Modernizing Bioeconomies*. On the enrolment of soil microbes in efforts to achieve sustainable agricultural intensification, see Krzywoszynska, “Nonhuman labor and the making of resources”; and Lyons, *Vital Decomposition*.
82 Battistoni, “Bringing in the work of nature,” 5.
83 Moore, *Capitalism in the Web of Life*, 2.
84 Battistoni, “Bringing in the work of nature,” 24.
85 Graeber, *Bullshit Jobs*, 238.
86 See Kareiva *et al.*, *Natural Capital*; and Helm, *Natural Capital*.

87 In Moishe Postone's terms, capital is "the central category of a society that becomes characterized by ... production for the sake of production, by a process that exists for the sake of process" (Postone, *Time, Labor, and Social Domination*, 269). To speak of natural capital does nothing by itself to call the pursuit of production for the sake of production into question.

88 "Nature-based solutions" are often defined explicitly as "solutions to societal challenges that involve *working* with nature" (Seddon *et al.*, "Getting the message right on nature-based solutions to climate change," 1518, emphasis added).

89 As Scott Kirsch and Don Mitchell have written, "dead labor" may be thought of as "work ossified and made concrete in the shape and form of a machine, a building, a finished commodity, a technological artifact, a piece of property, or even nature itself" (Kirsch and Mitchell, "The nature of things," 696).

90 In effect, the governance logic of modern bioenergy aims to replace the formal subsumption of nature's dead labour—fossil fuels—with the real subsumption of a much-expanded reserve of living labour, "invok[ing] the logic of subsumption on a planetary scale" (Carton *et al.*, "Revisiting the 'subsumption of nature'," 792).

91 Jonsson, "Abundance and scarcity in geological time."

92 Walker, "Keeping the lights on."

93 Daggett, *The Birth of Energy.*

94 Searchinger *et al.*, "Fixing a critical climate accounting error."

95 Malm, *Fossil Capital.*

96 Sedgewick, *Coffeeland*, 85.

97 Myers, "Photosynthetic mattering," 123.

98 Tyndall, *Heat Considered as a Mode of Motion*, 507.

99 Myers, "Photosynthetic mattering," 125.

2

The new biofuel rush: cars, crops, climate and coloniality

> At first sight [biofuels are] a beautiful idea. Instead of dragging a filthy black sludge from the bowels of the earth, we would grow our fuel in swaying fields of grain or flowers. ... There would be no more tankers running aground on rocky shores, no more corrupt regimes propped up by black gold, no more murders of indigenous people, no more wars in the Middle East.
>
> —George Monbiot, *Heat*, 2007

> Producing biofuels today is a crime against humanity.
>
> —Jean Ziegler, UN Special Rapporteur for the Right to Food, 2007[1]

Pull into a fuel station in a car, van, bus or truck almost anywhere across the UK, EU or US today, and you will very likely find it impossible to avoid purchasing biofuel. Petrol and diesel are now routinely sold not in pure form, as products derived entirely from fossil oil, but rather as blended fuels containing 5, 10 or sometimes even more than 20 per cent bioethanol or biodiesel, the vast majority of which is derived from plants. Whether processed from sugar- or starch-rich crops like sugarcane and corn, or alternatively from oil-rich crops like palm oil and soy, transport biofuels are now an unavoidable part of the reality of driving conventional internal combustion engine vehicles across an expanding portion of the globe. Well over 150 billion litres of transport biofuel are already consumed globally each year, with expert forecasts predicting an increase to more than 200 billion litres by 2028.[2] Even as recently as 2002, however, global transport biofuel consumption struggled to exceed 35 billion litres (measured on an oil-equivalent basis), of which around 50 per cent was concentrated in just one country—Brazil.[3] In North American and European contexts, the relative growth in the use of these fuels which has occurred since the turn of the century is thus difficult to overstate. Some have even called it a "new gold rush".[4]

Unless you pay particularly close attention to what you're putting into your car when you refuel it, however, you could be forgiven for overlooking the presence of biofuels altogether. Indeed, the majority of road transport biofuel in the global North is consumed remarkably unwittingly, with motorists being far more concerned by the prevailing price of forecourt petrol and diesel than by the changing composition of these fuels over time. In less than thirty years, biofuels have become an inescapable feature of modern transport systems in the global North not by battling to win hearts and minds, but effectively by stealth. The consequences of the radically increased production of these fuels, however, have been far less inconspicuous than the fuels themselves. Moreover, those consequences—which include the impacts of large-scale, intensive energy crop cultivation on food security, biodiversity and land rights—have been disproportionately concentrated into global regions located well away from the UK, EU and US, playing out instead across diverse landscapes in Latin America, South East Asia and sub-Saharan Africa. While biofuel itself quite literally blends into existing fuel infrastructures and cultures, then, to the point of being close to invisible to the drivers who consume it, the impacts that biofuel production has had in other places and on other people have been hard to deny and—for many—impossible not to vociferously oppose.

The central argument of this chapter is that this paradox—of transport biofuels as at once invisible to those who use them, and yet brutally consequential for societies and ecosystems located conveniently out of sight—is not an accident, but a central feature of the regulatory logic of modern bioenergy. Indeed, the new rush for biofuels of the early 2000s can be viewed, I argue, as a harbinger of the more fully developed modern bioenergy logics discussed in later chapters, which seek actively to reconfigure plant metabolism to boost the productivity and energy content of key energy crops, and moreover to enrol plant growth and burning as a basis for reconfiguring and stabilising the climate system. Set against these vaulting ambitions, the aim of earlier efforts to promote biofuel production—to harness a renewable alternative to petrol and diesel for fairly modest levels of use in road transport—might seem prosaic or even potentially innocuous. Nonetheless, driven by ambitious policy commitments made on both sides of the Atlantic, and ostensibly linked to growing concerns over soaring crude oil prices, climate change and rural development, the rapid global expansion in transport biofuel production which took place after the turn of the century can be viewed, I argue, as a watershed moment ushering in a distinct new geography of hydrocarbon provisioning—one with profound implications for power relationships shaping land ownership and use in territories extending far beyond the affluent global North.[5] In so far as early policies aiming to increase the use of biofuels brought about a "rush" of any kind, in fact, this was arguably not a rush towards some new, bio-based "gold", but a headlong rush for land itself.[6]

To unpack these arguments and illustrate the significance of early developments in the emergence of a new global economy of modern bioenergy, this chapter begins by situating moves to embrace plants as a substitute for petrol and diesel in the early 2000s within a longer historical context—one attuned to the contingency and fragility of close relationships between internal combustion engines and fossil fuels, which have all too often been recounted as inevitable. Drawing on expert interviews and documentary analysis, the chapter then moves on to scrutinise visions of a biofueled future animating policy debates in the UK, EU and US at the turn of the century, highlighting stark contradictions between aspirations for domestic feedstock production and fuel security on the one hand, and neoliberal trade frameworks governing international flows of agricultural commodities on the other. Following a brief overview of the litany of deleterious social and ecological consequences which have flowed from the rapid expansion of biofuel feedstock cultivation in diverse parts of Latin America, South East Asia and sub-Saharan Africa, the chapter ultimately advocates a view of transport biofuel policies implemented in the affluent global North as vehicles serving to renew and extend longstanding colonial logics of enclosure, violence and dispossession in the global South, all under the auspices of an approach to "picturing" land itself as idle, marginal or simply empty.[7] In short, far from constituting a "win-win-win" (or "silver bullet") capable of tackling climate change, energy insecurity and rural poverty in one fell swoop,[8] the new rush for biofuels of the early 2000s might be better seen—I argue—as having laid the foundations for a much broader and more deliberate neocolonial project of resource extraction and appropriation, one whose ultimate goal today is to make plant matter the world over into an energy resource for the already affluent, rather than into a subsistence resource for the majority.

From peanuts to panacea: the not-so-new biofuel rush

The idea of using plants and other biological feedstocks to produce transport fuel is not new. In fact, biofuels have a history as long-running as that of the internal combustion engine itself. At the outset of the twentieth century, the German mechanical engineer Rudolf Diesel demonstrated a small prototype of his own version of the internal combustion engine at the Exposition Universelle in Paris. The engine was powered not—as one might expect—by diesel, but by peanut oil. Whilst Diesel designed his engine to run principally on fossil oil, it was nonetheless his view that the use of peanut or other vegetable oils to propel motor vehicles could easily become commonplace.[9] In this respect his vision was also shared by Henry Ford; whilst buyers of

Ford's famous Model T may have faced a restricted choice over the colour of their vehicle in the 1910s,[10] they did at least have the option of fuelling their journeys using bioethanol, petrol or a combination of both. Consequently, whilst it seems entirely natural today that the majority of the world's road transport vehicles should be propelled by fossil fuels, in the early decades of the twentieth century the matter was far less clear cut. In the American context, as Jeffrey T. Manuel has shown, the appeal of deriving transport fuels from agricultural resources derived at least partly from public concerns about a "powerful and much-hated monopoly" controlling the oil sector, coupled to a desire to support "the venerated American farmer".[11] By the 1930s, however, the greater energy density of conventional gasoline, coupled with the discovery of vast new oil fields in several US states, had tipped the scales firmly in favour of fossil-based forms of vehicle propulsion.

Throughout the remainder of the twentieth century, whilst the industrial-scale conversion of biomass resources into transport fuel remained only a marginal activity in the US and other rich nations,[12] it did become prominent elsewhere. Most notable in this respect is the case of Brazil, where the mass production of bioethanol, usually derived from sugarcane, has a history dating back half a century.[13] Driven by fears over energy security in the wake of oil price shocks of the early 1970s, Brazil's military government instituted a Pro-Alcohol Programme in 1975, with the aim of meeting 20–25 per cent of the country's road transport fuel needs through the use of domestic biomass resources. By the mid-1980s, around 12 billion litres of bioethanol were being produced annually under this scheme, with bioethanol-dependent cars accounting for around 90 per cent of all new car sales in the country.[14] Brazil's success in integrating biofuel into its road transport infrastructure owed much, however, to the comparatively high price of oil during this turbulent economic period. From the late 1980s onwards, as oil prices began to fall once more, bioethanol production from sugarcane in the country declined steeply.[15]

By the beginning of the twenty-first century, when rising oil prices once again helped biofuels to re-emerge as an attractive economic proposition in Brazil, they had also been thrust onto the energy policy agendas of countries across much of the affluent global North as well. In the US, for example, a gargantuan objective of quadrupling bioethanol and biodiesel use by 2022 was enshrined in the Energy Independence and Security Act of 2007, building on more modest targets contained in the 2005 Energy Policy Act.[16] Throughout the subsequent two decades, energy security concerns—and especially concerns about a growing dependence on oil imports from the Middle East—have been by far the most prominent reason cited by advocates for expanded domestic biofuel production in the US.[17] Some commentators, as Michael Charles and colleagues have documented, even went as far as to frame the consumption of biofuels as "a small step that the average American can take in winning the war against terror".[18] Yet the New York terror attacks of 11 September 2001 did not give rise to a security-oriented energy policy discourse on their own. Rather, they served to amplify well-established calls for a diversification of the American fuel supply which had been kickstarted just before the turn of the millennium by spiking oil prices.[19] Against this fractious backdrop, then-President George W. Bush, on signing the Energy Policy Act into law in August 2005, offered a clear vision of "home-grown" biofuel as a means of shoring up American energy security, delivered in his inimitable style:

> I really wasn't kidding when I said, some day a President is going to pick up the crop report—(laughter)—and they're going to say we're growing a lot of corn, and—or soybeans—and the first thing that's going to pop in the President's mind is, "we're less dependent on foreign sources of energy". It makes sense to promote ethanol and biodiesel. (Applause.)[20]

Politically potent as these arguments undoubtedly were at the time, however, neither rising oil prices, nor even emotive arguments about a campaign to help freedom and democracy triumph over foreign terrorists, can fully explain the stunning resurgence of political interest in biofuels at the turn of this century. In the EU in 2001, for example, any boon that biofuels promised to yield for security of energy supply was set on equal terms with significant hopes for climate change mitigation and—crucially—rural development. As an influential European Commission White Paper of the time gushed, "[T]he spread of biofuels will help to reduce the European Union's energy dependency, improve the environment and also diversify production and jobs in agriculture".[21] Two years later, the EU passed into law the first of several policies enshrining indicative targets for the increased use of biofuels in road vehicles, the Biofuels Directive.[22] In 2009, this Directive was supplanted by legislation introducing binding targets for all EU member states to derive 10 per cent of their transport fuel from renewable resources, principally biofuel, by 2020.[23] Meanwhile, when the UK announced plans for its own specific legislation mandating the blending of biofuels into petrol and diesel, in late 2005, then-Secretary of State for Transport Alistair Darling put environmental questions—and especially climate change—front and centre of the government's rationale, ahead of both matters of energy security and indeed support for farmers.[24] The following summer, then-Prime Minister Tony Blair even found time to attend in person the opening of a £45 million facility turning oils from rapeseed, palm and soy into biodiesel in the north-east of England, where he talked up the capacity of biofuels to help the UK meet its GHG emissions reduction targets under the Kyoto Protocol.[25]

Between 2000 and 2005, then, biofuels appeared in a number of affluent countries as "an idea whose time had come".[26] An imaginary of biofuels as something of a "silver bullet"—a solution at once to all three pressing challenges of climate change, energy security *and* rural development—was commonplace, especially in the UK and European contexts.[27] Governments, moreover, were not alone in their enthusiasm for this idea. In addition to appealing to car manufacturers and potential investors in a nascent biofuel industry, many environmental groups were also on board. In a letter published in *The Guardian* in June 2004, for example, written against a backdrop of growing societal unrest over rising fuel prices in the UK, no fewer than five prominent environmental campaign organisations—the Green Alliance, Transport 2000 (now known as the Campaign for Better Transport), Friends of the Earth, the World Wildlife Fund and the Royal Society for the Protection of Birds—explicitly called on the government to "encourage the use of biofuels".[28] By 2011, at least seventeen separate countries had introduced legislation specifying targets for the increased blending of biofuels into road transport fuel.[29] Many governments had also complemented these targets with fiscal measures to kickstart the biofuel industry, including tax credits and subsidies for firms investing in bioethanol or biodiesel production. Despite this expansion, however, it has also been clear to a wide array of commentators since at least the late 2000s that a new rush towards biofuels did not—despite George W. Bush's assurances—really "make sense". Indeed, that rush has long been described as a form of "irrational exuberance" and even as the potential cause of "a global humanitarian disaster".[30] What, then, had gone wrong?

The remainder of this chapter focuses on the underlying contradictions of biofuel policies as introduced in the early 2000s by governments in the US and Europe. Significantly expanded demand for liquid transport biofuels has not emanated solely from these two locations over the past two decades, of course. Indeed, as of 2023, biofuel blending mandates were in force in no less than sixty countries (see Figure 2.1),[31] while increased consumption over the coming years is expected to be driven principally by Brazil, India, Indonesia and Malaysia.[32] Nonetheless, this chapter's focus on the US and EU can be justified not only by the fact that these players were both influential first movers in the new biofuel rush, but also because they remain—despite considerable controversy—two of the world's largest markets for these fuels today, together accounting for 56 per cent of global consumption, or around 190 million litres per day when measured on an oil-equivalent basis, in 2023.[33] Even in supply terms, while Brazil remains a bigger player than Europe—the former accounting for 22 per cent of global liquid transport biofuel production in 2023, as compared to the latter's 15 per cent—output in the US remains highest of all at 39 per cent, or close

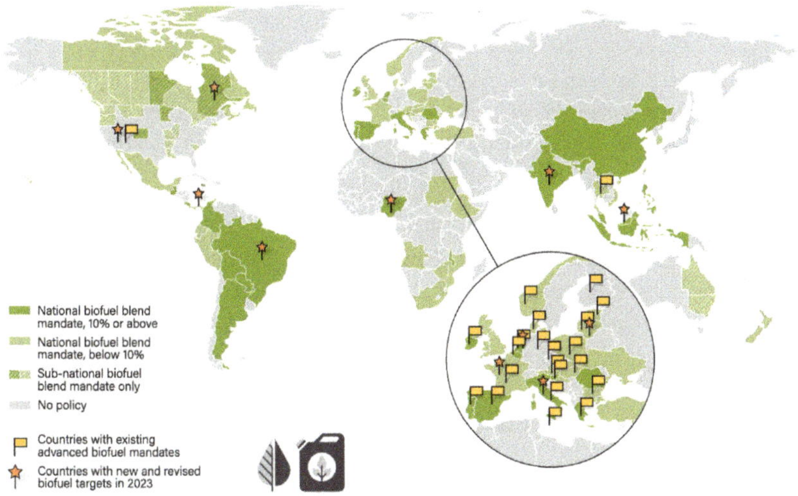

Figure 2.1 Countries with active biofuel blending mandates as of 2023. *Source:* Renewable Energy Policy Network, *Renewables 2024 Global Status Report*, p. 46.

to 130 million litres of oil equivalent per day. By contrast, while it may be a world leader in solar power, China accounted for just 12 million litres of liquid transport biofuel production per day in 2023, or around 4 per cent of global supply. The largest biofuel sector in Asia, by some distance, can instead be found in Indonesia, where 2023 production outstripped that achieved in China by a factor of 2.5.[34]

Biofuels, land and global trade: losing control or ignoring the inevitable?

In October 2006, experts affiliated with two consultancy outfits specialising in environmental law and global trade policy published an esoteric discussion paper examining what they termed "opportunities and constraints in the creation of a global marketplace" for biofuels. Some of the paper's opening lines have turned out to be extremely prescient:

> Producing fuel from agricultural crops has already raised questions about the impacts on the supply of food and land. What has received less attention is the shift that would result in the location of feedstock and fuel production. Energy demand and land productivity are somewhat asymmetrical. OECD [Organisation for Economic Co-operation and Development] countries, which

> import most of their fossil fuel, consume more than 49 million barrels of oil a day. While their demand for biofuels is therefore rising, the same industrialized countries do not have sufficient land availability to entirely meet that demand with domestic production. ... The most ideal land for sugarcane and oil palm trees, currently the most energy efficient *biofuel* feedstocks, is primarily located in developing countries in tropical and sub-tropical climates.[35]

It seems obvious in hindsight that policy commitments to replace any meaningful volume of fossil fuels with biofuels in the transport sector would inevitably have implications for global land-use change. At the outset of the century, however, governments in the US and Europe alike were far more concerned with unleashing the perceived untapped potential of domestic biofuel production than they were with assessing the potential impacts of biofuel mandates on food production, habitat loss or land rights abroad. In the US, notably, levels of bioethanol production—mostly derived from corn—were already considerable even prior to the formal endorsement of a new rush for biofuels by the 2005 Energy Policy Act. By 2006, the US had become the most significant bioethanol producer globally, registering a total output of more than 18 billion litres, exceeding even Brazil's annual sugarcane-based bioethanol production of the time. For comparison, total production of all biofuels—both bioethanol and biodiesel—achieved across the EU in the same year barely exceeded 6 billion litres.[36] In pursuing dreams of significantly increased energy independence in the context of volatile oil prices, however, the Bush administration's 2007 proposal to rapidly scale up the US biofuel sector, aiming for a lofty target of just over 130 billion litres of "home-grown" alternative fuels by 2022, nonetheless made production in 2005 look rather modest.[37] While the "serious challenge of climate change" was not left entirely unacknowledged by Bush in his justification for legislating such a strenuous mandate, enabling Americans to enjoy "lives less dependent on oil" was clearly positioned as the primary objective.[38] Lurking under the surface of all of these claims, however, was an equally pressing awareness of the appeal of biofuels to America's agricultural sector. As one leading scientist, a co-author of an influential 2009 International Council for Science report on the environmental and land-use consequences of biofuels,[39] remarked of the US political scene at this time:

> I think, you know a lot of politicians became committed to biofuels before there was much scientific analysis—there was some but not a lot. And it seems like a good idea right? It's green, it's renewable, you get rid of that nasty oil stuff, and you've got something that helps farmers out—it's one of these "win-win" situations. ... And corn growers of the US are a very powerful lobby, and they're tied into some very powerful corporations—it's hard to counteract that.[40]

From this standpoint, the enthusiasm exhibited by US policymakers for radically expanding domestic biofuel production in the early to mid-2000s can be best explained not by recourse to comprehensive assessments of the likely benefits of such a move for the climate, for energy security or even for farmers per se, but rather by an intuitive sense that any proposal with so many potential upsides simply could not go wrong. Yet, to borrow the terms of a more recent retrospective assessment, however "well-intentioned" these proposals may have been, if the plan was indeed "to help the climate by helping American farmers", pursuing this dream has ultimately come at a "heavy price", with consequences felt not just within the US but also by communities and ecosystems located—as will be discussed later—well beyond America's own borders.[41]

In the EU in the early 2000s, meanwhile, even if rhetorical imbalances between visions of enhanced energy security and aspirations to tackle climate change may have been the mirror image of those evident in US justifications for a biofuel mandate, an underlying desire ultimately to solve problems relating to agricultural policy is also strikingly evident in hindsight. It is certainly no coincidence, for example, that the EU's flagship biofuels policy was passed into law in May 2003, just one month prior to the adoption of unprecedented reforms to the bloc's Common Agricultural Policy (CAP), wherein support payments provided to farmers were for the first time effectively delinked from the growth of specific crops and replaced with a so-called "Single Payment Scheme" for all land kept in "good agricultural condition", regardless of whether that land was to be used for active cultivation or not.[42] Having raised the prospect of subsidising farmers for producing nothing, the EU's commitment to ramp up biofuel production was influential in staving off critics of these 2003 CAP reforms. Indeed, the text of the Biofuels Directive itself does not refrain from highlighting these aspects of the policy, noting that "promoting the use of biofuels ... could create new opportunities for sustainable rural development in a more market-oriented common agricultural policy geared more to the European market".[43] In an interview, one European Commission official close to biofuels policy deliberations in Brussels for several years put things more directly:

> The biofuels policy in a nutshell I would say was put in place to basically rescue the Common Agricultural Policy. And it was there because you had, basically lots of surplus land, you were paying farmers basically to do nothing, and somebody had the bright idea: "OK, you could grow energy crops on this, it would use the land, it would give the farmers some income, and you know we could say it gives us GHG savings, energy security benefits and all those sorts of things".[44]

For distinct reasons in each case, then, policymakers in the US and the EU alike were both resoundingly focused, in the early 2000s, on rationalising commitments to biofuel mandates which—for all their outward talk of reducing GHG emissions and increasing energy independence—were remarkably inward looking: focused no more on global environmental or geopolitical challenges than on resolving political problems linked firmly to the realm of domestic agriculture.[45]

The question of whether policymakers on either side of the Atlantic genuinely believed that it would be possible to meet ambitious targets for scaled-up biofuel use entirely on the basis of domestic crop production is, of course, impossible to answer definitively. As John F. Kennedy once remarked, "[T]he essence of ultimate decision remains impenetrable to the observer—often, indeed, to the decider himself".[46] What it is possible to say with certainty, however, is that expert analysts at bodies like the International Energy Agency had—from an early stage—been noting "a mismatch" between the main geographical locations of rapidly rising biofuel demand and those regions, typically in the tropics and sub-tropics, where biofuels could be produced most efficiently and cheaply.[47] In the UK, the Department for Transport (DfT) commissioned an independent expert review of the potential "global impacts" of road transport biofuels as early as 2004—not only did this report question whether "the UK could sensibly aim to be self-sufficient in transport biofuels", but it also suggested that an exclusively home-grown approach was "unlikely ... in practice anyway, as imports are likely to occur".[48] To some extent, this conclusion was based on estimates of the area of land likely to be required to meet mandates from domestic production alone. Writing in the same year as the DfT-commissioned review, the International Energy Agency had estimated that replacing 10 per cent of petrol and diesel in the US by 2020 would require no less than 43 per cent of existing cropland to be dedicated to biofuel feedstock cultivation; in the EU, the corresponding figure for meeting an identical target was 38 per cent of existing cropland.[49] In the UK, a report examining the "prospects and challenges" of sustainable biofuels issued by the Royal Society—the country's pre-eminent scientific academy—had indicated that meeting a blending target of just 5 per cent would still require between 10 and 45 per cent of all cropland to be dedicated to biofuel feedstock cultivation.[50]

Beyond the likely modest yield potentials of domestic biofuel feedstock cultivation in the US and EU, however, dreams of genuine self-sufficiency in biofuels were ultimately to be scuppered by an even more inconvenient truth—under international trade laws set out by the World Trade Organization (WTO), banning imports of biofuel feedstocks altogether would simply be impossible. Indeed, while the US and EU both retained appreciable tariffs on imports of bioethanol well into the 2000s, affording domestic producers a relative degree of immunity to international competition, cross-border trade in biofuels nonetheless expanded rapidly from the mid-2000s onwards. By 2008, for example, an estimated 1.9 billion litres of bioethanol were being imported to the US to make up a growing shortfall between domestic supply and demand; by 2016, imports of biodiesel were running in the order of close to 400 million litres per *month*.[51] In the EU, historically a global leader in biodiesel production, domestic supply was initially able to keep pace with demand, but this gave rise to a precarious situation in which "no or low tariffs" were applied to imports of biodiesel in most member states.[52] As overall EU demand for biodiesel grew steadily through the end of the 2000s and into the 2010s, therefore, the bloc's biofuel mandate inevitably began to give rise (despite having been put in place allegedly to "rescue the Common Agricultural Policy"[53]) to growing imports of feedstocks like palm oil and soybeans, particularly from tropical regions where "suitable and available land is mostly concentrated, and annual yields are highest".[54] As one representative of a prominent environmental campaign organisation which had been critical of biofuel mandates in Europe at the time remarked, bluntly, in an interview: "the farmers thought this was going to be a goldmine, which of course it isn't, unless there's some protectionist measures put in place, because stuff tends to be done cheaper outside Europe".[55]

By 2008, biofuel mandates—far from serving to reduce transport GHG emissions and oil dependency through increased domestic agricultural output alone—were instead widely being called into question over their potential to facilitate "the 'export' of environmental problems to developing regions".[56] From a standpoint concerned principally with climate change and biodiversity, for example, enthusiasm had been severely dampened by links with deforestation, and indeed other forms of damaging land-use change, in countries like Indonesia and Brazil, from where feedstocks like palm oil, soy and sugarcane were being sourced to plug shortfalls in domestic bioethanol and biodiesel supply from home-grown crops like corn, wheat and oilseed rape.[57] From a humanitarian standpoint, meanwhile, biofuels had also been forcefully described as a "crime against humanity" by a UN Special Rapporteur on the Right to Food.[58] Indeed, in light of their inherent requirement for agricultural land, biofuels were by 2008 routinely being invoked as key drivers not just of rising global prices for various staple food crops, including in a testimony provided by the Chief Economist of the US Department of Agriculture to a Joint Economic Committee of Congress,[59] but also of growing instances of rural communities being dispossessed from their land across many parts of the developing world.[60] Having unleashed such a diverse array of "unintended consequences",[61] policymakers could no longer live "in denial";[62] the EU's Environment Commissioner, for instance, was forced to admit that problems with biofuel production were "bigger than we thought", raising the prospect that blending targets might simply have to be missed to avoid doing further harm.[63] In the UK, the government's foremost scientific adviser on environmental policy even went as far as to suggest that it would be "insane" to press ahead with biofuel mandates if the sustainability of biofuel production itself could not be guaranteed.[64] In short, in the space of just a handful of years, biofuel blending mandates had become perhaps the most vehemently opposed—and satirically derided (see Figure 2.2)—environmental laws to be found in both the US and the EU. Despite the turmoil of this controversial period, however, biofuel blending mandates were ultimately retained as key aspects of renewable energy policy on both sides of the Atlantic. As the next section will show, the arguments which were used to steady the biofuels ship in the late 2000s have had far-reaching and long-lasting consequences for modern bioenergy logics writ large.

Figure 2.2 A satirical cartoon published in 2015, implying links between official EU biofuel blending mandates and deforestation.
Source: Reproduced with permission from original copyright holder Andy Davey.

Biofuels and indirect land-use change: admit defeat or double down?

If early controversies around efforts to promote transport biofuel use in the US and EU could potentially be diffused—in theory if not in practice—by an insistence that blending targets would be met solely through domestic feedstock cultivation, from 2008 onwards the distinction between home-grown and imported biofuels was becoming increasingly blurred. The idea of importing fuels produced on land which had previously played host to rainforest, or otherwise ecologically valuable landscapes, had never attracted many explicit advocates, of course. In early 2008, Professor Sir John Beddington, then newly appointed as the UK Government's Chief Scientific Adviser, went on record to label this practice "profoundly stupid".[65] Scientific assessments, meanwhile, suggested that the GHG emissions associated with clearing rainforest to make way for soybean cultivation in Brazil could take more than three centuries to offset by subsequently substituting conventional diesel fuel with the resulting biodiesel product.[66] The more thorny policy paradox, however, was one to which even biofuels grown *inside* the US and EU were not immune: put simply, when land already playing host to agriculture was diverted to grow feedstocks for biofuels, what happened to the food production that had been displaced? Where did this production go? Might not these so-called "indirect" land-use change (ILUC) effects potentially also wipe out any environmental benefits associated with home-grown biofuels?

In early 2008, the journal *Science* published two momentous articles, over successive pages, reporting estimates of the volume of GHG emissions generated when land is converted to agriculture to help satisfy rising demand for biofuels.[67] Crucially, the second of these two articles focused not on direct land-use change, but rather on landscape clearances promulgated indirectly through global agricultural markets. The lead author of this second article, Timothy Searchinger, a US-based environmental lawyer working for the Environmental Defense Fund, had allegedly been alerted to the potential risks of ILUC while listening to then-President George W. Bush outline ambitious plans to rapidly expand domestic bioethanol production in his 2007 State of the Union address.[68] For Searchinger, Bush's plans—including a headline commitment to replace around 130 billion litres of gasoline and diesel with home-grown alternative fuels by 2022—simply didn't add up. Too much of America's cropland was already under intensive production, and too little arable land was otherwise unused, for such a vast additional volume of feedstock to be derived domestically. Using a worldwide agricultural model, Searchinger and his colleagues estimated that corn grown for bioethanol production on American croplands would—by virtue of displacing millions of hectares of food production to locations like Brazil,

China and India—generate enough GHG emissions from land-use change to render that bioethanol twice as damaging for the climate over a thirty-year period than simply continuing to burn fossil-derived gasoline.[69]

While Searchinger and his team struggled to turn the tide of biofuel policy in the US with their analysis, their findings on ILUC eventually made significant waves across the Atlantic. Policymakers in the EU and UK both commissioned substantial expert reports into the indirect impacts of biofuel production in the late 2000s, although at the EU level especially, there was initially some resistance to acknowledging the potential scale of the issue at hand. Indeed, despite policy officials in Brussels having been made aware of Searchinger's findings in late 2007,[70] when the European Commission set out proposals to increase targets for renewable transport fuel use in early 2008, an official impact assessment document, itself published just two days before Searchinger's own paper appeared in *Science* in February 2008, dismissed the ILUC problem as follows:

> Some commentators imply that each hectare of land devoted to the cultivation of crops for the biofuel market will have to be offset by finding a hectare of land, somewhere else, to produce the food that would (it is suggested) otherwise have been produced on the biofuel-producing land. However ... there is a clear link between demand for agricultural commodities, their prices, and investment in agriculture and agricultural productivity. Thus, it can be expected that the main impact of increased biofuel demand will be a further increase in productivity, not an increase in the quantity of land use for agriculture.[71]

While at least some Brussels bureaucrats may have been reluctant to open the can of worms presented by ILUC, apparently believing that crop yield increases would offset any putative requirement for additional land to meet biofuel mandates, policymakers in the UK reacted more decisively. Just eight days prior to the publication of Searchinger's analysis, Ruth Kelly—then UK Secretary of State for Transport—ordered an independent review into the "indirect effects of biofuels production", noting that "we are not prepared to go beyond current UK target levels for biofuels until we are satisfied it can be done sustainably".[72] Then, just a month later, the European Commission's in-house research arm—the Joint Research Centre (JRC)—published its own report into biofuels, effectively forcing the hand of policymakers in Brussels as well, by concluding that: "The uncertainties of the emissions due to indirect effects, much of which would occur outside the EU, mean that it is impossible to say with certainty that the net GHG effects of the biofuels programme would be positive".[73]

Beyond compounding the sombre message which had been conveyed by Searchinger and his team just a month earlier, the JRC's report served to significantly ramp up pressure on EU policymakers, laying bare that even the European Commission's own scientists recognised ILUC as an issue which could easily negate, or even outweigh, the climate benefits that biofuels would otherwise have been able to deliver. In so doing, the report ushered in an era of intensifying scrutiny in Brussels, with concerns being articulated by a growing array of campaign organisations and expert bodies,[74] and internal dialogue between the transport, climate and agriculture branches of the European Commission becoming increasingly acrimonious.[75] Ultimately, by the time increased EU targets for biofuel use were formalised in April 2009, under the guise of the Renewable Energy Directive, concerns about ILUC had solidified to the extent that policymakers were bound, by this new legislation, to produce a report "reviewing the impact of indirect land-use change on greenhouse gas emissions and addressing ways to minimise that impact".[76]

If policymakers in the UK and especially the EU were reluctant to take immediate action to change course on biofuels in the face of the ILUC problem, much of that reluctance might ultimately be traced to an acute awareness of the interests of a new and increasingly influential economic sector—the biofuel production industry—whose very existence could in large part be traced directly to those same policymakers' enthusiastic embrace of biofuels only half a decade earlier. As one expert scientist with first-hand experience of policy debates taking place in Brussels explained:

> There's a big biofuels industry that's now appeared, which relies for its entire existence on public subsidy and puts a vast amount of money into lobbying. And of course they're not as big as the oil industry or whatever, but they're much more ... they put a much bigger percentage of their resources into lobbying.[77]

Similarly, for a representative of one prominent environmental campaign group active in debates in Brussels at the time: "This [biofuels] industry didn't exist a few years ago, and now they can pay for an army of lobbyists, and buy scientists to do their own domestic research".[78]

Not only were the biofuel industry's lobbying activities in Brussels highly visible in the period immediately following the emergence of ILUC as a potential problem; in many cases, their activities were also strategically and commercially aligned with much larger and more economically powerful conventional oil companies and groups. The implications of both of these facts were well articulated in the words of a spokesperson for the European Environment Agency:

> They've [the European Commission] built up an industry, which they are kind of reluctant to kind of ... drop, in a way, and obviously there's now really big players involved like BP, Shell and so on. They all have interests now, they're really lobbying very heavily, I mean—when I go to these kind of stakeholder

> meetings [in Brussels], it's—that's completely dominated by industry now and they have very strong interests and very strong lobbying power.[79]

Put simply, then, a significant portion of the European Commission's reticence on ILUC might be explained by an awareness that any reduction in political commitment to biofuels could easily have resulted in legal action from the biofuels industry itself. Even in the UK, where policymakers had been relatively quick to commission an independent review into the indirect effects of biofuel mandates, that review's overall recommendations—published in July 2008—called not for any fundamental rethink of support for increased biofuel use in road transport, but rather for a less ambitious growth trajectory, serving not only to grant policymakers more time to gather evidence about how best to meet existing targets, but also to provide more certainty to businesses about the long-term future of the biofuel sector.[80]

In seeking to justify ongoing political support for increased biofuel use despite the potentially vast GHG impacts of indirect land-use change, protagonists of biofuels in the late 2000s and early 2010s tended to deploy three key sets of arguments. Firstly, in line with ideas put forward in the European Commission's own official impact assessment of its 2009 Renewable Energy Directive,[81] that growing demand for biofuels would inevitably drive increases in the efficiency and productivity of agriculture on existing land, such that not all of this demand would need to be met by expanding the footprint of farming into new terrain. Secondly, that there are always significant quantities of productive but abandoned agricultural land available globally, and that such "idle" lands could easily be appropriated for biofuel production without bringing about any indirect land-use change at all.[82] Finally, that scientific research would gradually enable biofuels to be derived from so-called "second-generation" or "advanced" feedstocks, which would not need to compete directly with food production—for instance by making use of the inedible, fibrous portions of staple food crops, or else by relying on the cultivation of dedicated energy crops, such as *Miscanthus* or *Jatropha*, on so-called "marginal" or "degraded" lands.[83] On the first count, while the idea of a strong positive link between increased demand for crops and overall yields may be logically compelling, evidence shows that the vast crop yield increases achieved globally between 1961 and 2005 occurred even as average prices declined by half in real terms.[84] These arguments also tend to overlook the potential additional GHG emissions that would be associated with further intensification, not least if heightened yields are achieved through the increased application of fertilisers. On the second count, the idea that vast areas of abandoned agricultural land could be straightforwardly brought back into production for biofuels founders on ambiguities around what types of land count as "abandoned"—not all land earmarked in this way might prove capable of sustaining commercially viable yields of energy crops, for example, especially if it were previously used for animal grazing. Growing energy crops on "idle" lands might not always evade opportunity costs, either; indeed, such land might well reforest naturally if left alone, thereby contributing to enhanced carbon sequestration from the atmosphere without biofuels needing to be produced at all.[85]

It is around the final set of claims marshalled by biofuel advocates in this period, however, of the potential for large volumes of second-generation or advanced biofuels to be grown on marginal or degraded lands, that perhaps the most pernicious set of ideas and assumptions can be found. Even on a discursive level, the use of the label "second-generation" conveys a view of biofuels not as hydrocarbon products arising in vivo, from agricultural activities embedded in a diverse range of socioecological contexts, but rather as isolated technologies whose efficiency and sustainability can be tailored and optimised in vitro, through the application of scientific and technical expertise. As Magdalena Kuchler has written, "[S]ociotechnical imaginaries of second-generation bioenergy technology ... project the illusion of utopian potential".[86] In so doing, the second-generation label implicitly endorses attempts to convert more and more biomass into combustible liquid fuels as inherently virtuous, regardless of the broader political context within which demand for biofuels is situated. As one bioenergy and sustainability expert working in the field of chemical engineering put it in an interview:

> Everyone is saying "oh this is a terrifically exciting scientific and technical challenge, how can we make second generation biofuels efficiently?" and no one—very few people at least—stop and say, "yeah but, should we be doing it anyway?" Even if you come out with something that is thermodynamically efficient, it still doesn't mean it['s] going to be worth doing. So there's a sort of naive technological enthusiasm for it.[87]

More concerning still than this "naive technological enthusiasm", however, are sweeping gestures towards marginal or degraded lands as an automatically more sustainable basis for growing biofuel feedstocks than lands already under intensive crop cultivation. The EU's 2009 Renewable Energy Directive, for example, states explicitly that "biofuels should be promoted in a manner that encourages greater productivity *and* the use of degraded land".[88] Perhaps unsurprisingly however, given the "vague" nature of the label, estimates of the globally available quantity of "degraded land" vary widely—from less than 1 billion hectares to over 6 billion hectares.[89] As critical scholars like Jennifer Baka and Jesse Goldstein have noted, imaginaries of marginal or degraded land frequently revive colonial logics by invoking the idea of "wastelands"—lands deemed mismanaged when set against capitalist logics of efficiency and productivity, and therefore in need of commercial investment and "improvement".[90] These colonial logics not only overlook the "high level and wide range of uses"—such as subsistence agriculture and pastoralism—already being made of such lands by their typically poor inhabitants.[91] In addition, they also frequently serve to legitimise formal territorial enclosures and the displacement of existing populations by powerful actors—whether states, multinational corporations or even multinational development organisations—all ostensibly under the guise of commitments to renewable energy transition and climate change mitigation.[92] Peter Dauvergne and Kate Neville, for example, identify "patterns of exclusion" arising from commercial biofuel development in the tropics, linked especially "to the designation of land as marginal or idle".[93] Sonja Vermeulen and Lorenzo Cotula, meanwhile, have identified persistent shortcomings in community consultation processes preceding the acquisition of large tranches of land for biofuel plantations in a range of African countries, including Mozambique, Tanzania and Mali.[94] Separately again, assessments of corporate-led energy crop cultivation based on the supposedly high-yielding, so-called "wonder crop"[95] *Jatropha* in Tamil Nadu, India, have found not only that such regimes ultimately yield less "useful energy" than prevailing fuelwood economies serving households and local industrial consumers, but also that the establishment of *Jatropha* crops frequently proceeds hand-in-hand with "the enclosure of common property lands".[96] Set against this backdrop, European and American political responses to intensifying controversies around the indirect effects of biofuel production in the late 2000s might be argued not merely to have succumbed to techno-optimism, but also to have turned a blind eye to the distinctly neocolonial "plunder of resources" from lands concentrated disproportionately in less affluent nations of the global South.[97]

Bio-fooled: from farming at home to sacrifice zones abroad

Despite being framed as a climate change mitigation tool and a key plank of the renewable energy transition, Europe and America's new rush towards biofuels in the early 2000s must be seen ultimately as a project which has reinforced, rather than disrupted, incumbent fossil fuel-powered transport infrastructures and industries. While the global biofuel sector has grown precipitously, from a daily total consumption of just under 34 million litres (measured on an oil-equivalent basis) in 2000 to almost exactly ten times this amount per day in 2023 (see Figure 2.3), this still facilitates the blending of only modest levels of plant-based fuel into conventional petrol and diesel, at concentrations of below 10 per cent in the vast majority of countries.[98]

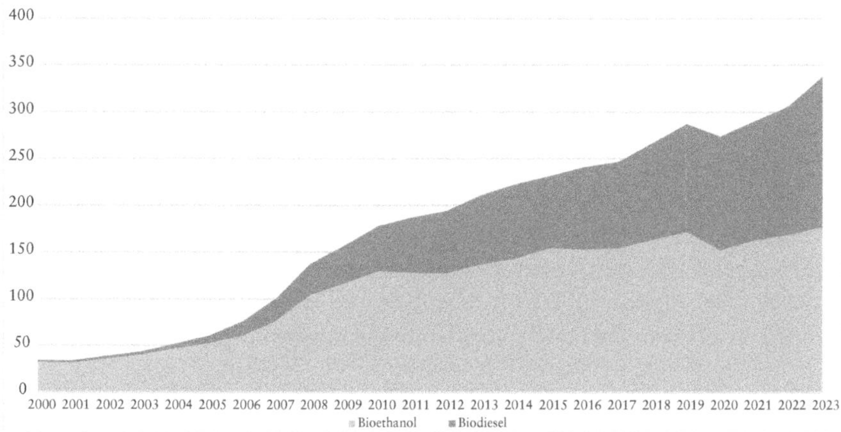

Figure 2.3 The evolution of global bioethanol and biodiesel consumption, 2000–23, in millions of oil-equivalent litres per day.
Source: Author's own graph, derived from data provided in the Energy Institute, *Statistical Review of World Energy 2024*, **p. 65.**

The new rush for biofuels, then, cannot strictly be understood as an attempt to fully replace or move away from fossil fuels. Indeed, viewed from a more critical perspective, it constitutes an archetypal manifestation of Christophe Bonneuil and Jean-Baptiste Fressoz's observation that "the history of energy is not one of transitions, but rather of successive *additions* of new sources of primary energy".[99] The low-level blending of biofuels may displace some fossil fuel burning on paper, of course. In material terms, however, the growth of the global biofuel sector has served to help fossil fuels remain the dominant source of propulsive power for road vehicles not just in affluent societies in Europe and America, but also in a growing array of developing country contexts too. To paraphrase Mark Simpson and Imre Szeman, the idea that there will ever be a wholesale "transition" to a fully biofueled future is thus best seen as a powerful "fiction"—in reality, biofuels are already integral to "a present that withholds ... any capacity to create a genuinely different energy future",[100]

Despite its ambiguous relationship to imaginaries of energy transition, however, the new biofuel rush has nonetheless profoundly reshaped geographies of global hydrocarbon fuel production, trade and consumption over the past two decades. Making a mockery of early political visions of European and American self-sufficiency in biofuels, even modest attempts to curtail the unfettered global trade of one key feedstock now deemed too environmentally damaging to help contribute to climate mitigation targets—palm oil—have precipitated protracted and acrimonious legal disputes at the WTO.[101] That the advance of official regulatory responses to the late-2000s controversy over ILUC has been so glacial and faltering, meanwhile, serves merely to underline that a turn to biofuels has not extricated Europe or America from the fractious resource geopolitics of oil; rather, this turn has simply introduced a new layer of hydrocarbon geopolitics—one involving major feedstock exporters like Indonesia and Malaysia—on top of long-standing international tensions centred around the vast fossil oil reserves of the Middle East. All the while, and in stark contrast to these slow geopolitical manoeuvrings, the actual transformation of swathes of Latin American, sub-Saharan Africa and South East Asia into biofuel plantations has proceeded at an eye-watering pace, sometimes involving large-scale acquisitions of hundreds of thousands of hectares of land at a time.[102] Whether these large-scale land acquisitions have been enacted legally or through a "murky politics of land grabs",[103] such moves have routinely been defended on the basis that an embrace of biofuels will both remediate and render productive lands currently languishing in a marginal, degraded, or otherwise abandoned state. A great twenty-first-century expansion in the spatial footprint of the global biofuel industry has thus been legitimised through distinctly colonial logics of "improvement"—logics envisioning lands as at once "underutilised" and "empty of people, histories, and claims", and yet at the same time "full of potential".[104] Whether by design or by accident, the logics of the new biofuel rush, a rush kickstarted by policy decisions taken in Brussels and Washington, DC, in the early 2000s, have thus had consequences—violence, dispossession and exclusion—hardly distinguishable from those which accompanied previous rounds of colonial resource extraction.

Contemporary assessments of the new biofuel rush have often positioned these developments as part of a wider "scramble" for farmland witnessed in the mid-2000s, driven initially by rising global prices for both food and oil, and then accentuated by the global financial crisis of 2007–2008.[105] More recently, as noted in the previous chapter, scholarship on the changing geographies of global energy provisioning has situated biofuel production within a broader "return to a land-based surface energy regime", one witnessing "enormous" investment in land for the development of renewable energy infrastructures of all kinds.[106] Crucially of course, and as Max Haiven has recently pointed out, there is nothing inherent to a plant like palm oil that serves to dictate "the context and methods in which it is manufactured and the functions it serves within a broader world capitalist system".[107] Nonetheless, my contention in this book is that the new biofuel rush of the early 2000s must be seen as having laid the foundations for a more fully developed regulatory logic of modern bioenergy—a logic which is distinct from wider global dynamics of both land speculation and acquisition, and indeed renewable energy transition. A key feature of this logic is its commitment to a specific "spatial ordering" of biofuel production, or what Arthur Mol has termed a "global biofuel scape", in which key centres of biofuel consumption in the global North are served not exclusively by national or regional programmes of biofuel production, but also by large-scale imports of an array of cheap feedstocks cultivated intensively in various parts of the global South.[108] For the global North citizens who ultimately consume these biofuels, this spatial ordering helps to preserve their relatively inconspicuous nature—already all but invisible at the point of use—by confining the most deleterious social and environmental effects of their production to distant lands located comfortably out of sight. This is thus a global biofuel "scape" predicated on a specific "production of space", one in which harms are concentrated disproportionately into "some places in order to protect the environmental health and sustainability of other places".[109] Particularly where intensive biofuel production takes place on lands which have had to be actively cleared—of vulnerable people and communities, of mature and highly biodiverse ecosystems and habitats, or both—it is no stretch to describe those lands as "sacrifice zones" deliberately surrendered for the supposed "greater good" of sustaining high-consumption energy systems and lifestyles in the global North, fuelled now by plants rather than fossils.[110]

While today's regulatory logic of modern bioenergy continues to exhibit and reinforce the highly uneven global spatial ordering of biomass resource provisioning which was unleashed by the new biofuel rush of the early 2000s, that logic is no longer merely geographical in scope. Indeed, scientific, commercial and governmental actors are increasingly turning to plants as a presumed basis for decoupling wider components of prevailing global energy systems from climate change, positioning biofuels as crucial to the future sustainability of electricity generation and even aviation. As later chapters will show, these expectations are giving rise not simply to a major spatial reordering of prevailing energy–society relationships, but also to efforts to optimise and improve crop plants on a molecular level, effectively producing new kinds of vegetal nature. Moreover, they are also giving rise to new political economies of bioenergy, wherein value is linked not just to the ability of crop plants to produce combustible fuels, but also to the deliberate enrolment of bioenergy into efforts to regulate the composition of the global atmosphere itself. Before moving on to scrutinise these aspects of contemporary modern bioenergy logics, however, the next chapter first turns its attention to the rapid emergence, since 2010, of a new frontier of biofuel production predicated on sacrifice zones located not in the tropics, but rather in the southern US, where the forest-based manufacture of wood pellet fuels has been integral to British and European efforts to replace coal and help facilitate so-called "net zero" electricity generation.

Notes

1 BBC News, "Biofuels 'crime against humanity'."
2 International Energy Agency, *Renewables 2023*, 94.
3 Energy Institute, *Statistical Review of World Energy 2024*, 65; see also International Energy Agency, *Biofuels for Transport*, 30.
4 Simpson *et al.*, "The new gold rush."
5 Bridge *et al.*, "Geographies of energy transition"; and Huber and McCarthy, "Beyond the subterranean energy regime?"
6 Scheidel and Sorman, "Energy transitions and the global land rush"; and Matondi *et al.*, *Biofuels, Land Grabbing and Food Security in Africa.*
7 Li, "What is land?"; and Baka, "Making space for energy."
8 Hitchner *et al.*, *Forests as Fuel*, 61; see also Palmer, *Science and Politics in European Energy and Environmental Policy.*
9 In 1912, Diesel remarked that "the use of vegetable oils for engine fuels may seem insignificant today but such oils may become, in the course of time, as important as petroleum and the coal-tar products of the present time."
10 I refer obliquely here to Henry Ford's famous declaration that "any customer can have a car painted any colour that he wants so long as it is black". See also Worldwatch Institute, *Biofuels for Transport*, 5.
11 Manuel, "Lessons from a forgotten fuel," 415; see also McCarthy, *Auto Mania.*
12 For an overview of modest efforts to promote the use of biofuels in Europe between 1900 and 2000, see Kutas, Lindberg and Steenblik, *Biofuels—At What Cost?* 15–17.
13 Eaglin, *Sweet Fuel.*
14 Wilkinson and Herrera, "Biofuels in Brazil."
15 Eaglin, "The demise of the Brazilian ethanol program."
16 Robbins, "Fuelling politics." In absolute terms, these targets sought to supply over 130 billion litres of biofuel to American motorists by 2022.
17 In January 2007, then-President George W. Bush spoke of a desire to "cut our total [oil] imports by the equivalent of three-quarters of all the oil we now import from the Middle East." The White House, "President Bush delivers State of the Union address."
18 Charles *et al.*, "Public policy and biofuels," 5740.
19 According to the International Monetary Fund, for instance, average crude oil prices had risen from a low of around $11 per barrel in February 1999 to peak at almost $35 per barrel in the first week of September 2000, dropping only slightly to remain at $32 per barrel throughout October and November of the same year. See International Monetary Fund, *The Impact of Higher Oil Prices on the Global Economy.*

20 The White House, "President signs Energy Policy Act."
21 European Commission, *European Transport Policy for 2010*, 87. For similarly positive policy appraisals of the time, see International Energy Agency, *Biofuels for Transport*; and UK Department of Trade and Industry, *Our Energy Future*, 69–70.
22 European Commission, "Directive 2003/30/EC."
23 European Commission, "Directive 2009/28/EC."
24 BBC News, "Green fuel obligation announced."
25 BBC News, "PM opens major biofuels complex."
26 While this aphorism is most frequently attributed to Victor Hugo, it chimes with work in political science scrutinising the influence of persuasive ideas over political agenda setting. See for example Kingdon, *Agendas, Alternatives and Public Policies*.
27 Oliveira *et al.*, "How biofuel policies backfire," 771. In the US, the climate change mitigation leg of this three-pronged argument was never articulated as strongly as elsewhere, mirroring a wider reluctance on the part of US policymakers to commit to tackling the problem, most obviously evident in George Bush's 2001 announcement that the country would not implement the Kyoto Protocol.
28 "Brown should stand firm on rising oil prices."
29 Robbins, "Fuelling politics."
30 Keeney, "Ethanol USA"; and Monbiot, *Heat*, 158.
31 Renewable Energy Policy Network, *Renewables 2024 Global Status Report*, 46.
32 International Energy Agency, *Renewables 2023*, 96.
33 According to the Energy Institute, a daily average of just over 337 million litres of liquid biofuels, measured on an oil-equivalent basis, was consumed globally in 2023. Of this total, consumption in the UK and Europe combined accounted for 18 per cent, while consumption in the US accounted for 38 per cent. See Energy Institute, *Statistical Review of World Energy 2024*, 65.
34 Energy Institute, *Statistical Review of World Energy 2024*, 64–65.
35 Howse *et al.*, *WTO Disciplines and Biofuels*, 3.
36 Worldwatch Institute, *Biofuels for Transport*, 6–7.
37 The White House, "President Bush delivers State of the Union address."
38 The White House, "President Bush delivers State of the Union address."
39 Howarth and Bringezu, *Biofuels*.
40 Interview, expert scientist, July 2010.
41 Lustgarten, "Palm oil was supposed to help save the planet."
42 Swinbank and Daugbjerg, "The 2003 CAP reform," 54.
43 European Commission, "Directive 2003/30/EC," 43.
44 Interview, European Commission official, June 2010.
45 In the EU context specifically, another interviewee went as far as to describe it as "nakedly obvious" that farmers had a "stranglehold" over Brussels at this time. See Palmer, "Stopping the unstoppable?" 998.
46 Allison, *Essence of Decision*.
47 International Energy Agency, *Biofuels for Transport*, 16.

48 Fergusson *et al.*, *Expert Paper on the Global Impacts of Road Transport Fuels*, 5.
49 International Energy Agency, *Biofuels for Transport*, 130.
50 Royal Society, *Sustainable Biofuels*, 53.
51 Worldwatch Institute, *U.S. Biofuels*; and Lustgarten, "Palm oil was supposed to help save the planet."
52 Worldwatch Institute, *Biofuels for Transport*, 143.
53 Interview, European Commission official, June 2010.
54 Doornbosch and Steenblik, *Biofuels: Is the Cure Worse Than the Disease?* 4.
55 Interview, environmental campaign group representative, July 2010.
56 Royal Society, *Sustainable Biofuels*, 63.
57 Royal Society, *Sustainable Biofuels*, 12.
58 BBC News, "Biofuels 'crime against humanity'."
59 Mitchell, "A note on rising food prices," 4.
60 Cotula *et al.*, *Fuelling Exclusion?*
61 Anderton and Palmer, "Evidence-based policy as iterative learning."
62 Giampietro and Mayumi, *The Biofuel Delusion*.
63 BBC News, "EU rethinks biofuels guidelines."
64 BBC News, "Call for delay to biofuels policy."
65 Randerson, "Food crisis will take hold before climate change."
66 Fargione *et al.*, "Land clearing and the biofuel carbon debt," 1237.
67 Fargione *et al.*, "Land clearing and the biofuel carbon debt"; and Searchinger *et al.*, "Use of U.S. croplands for biofuels increases greenhouse gases through emissions from land-use change."
68 Lustgarten, "Palm oil was supposed to help save the planet."
69 Searchinger *et al.*, "Use of U.S. croplands for biofuels increases greenhouse gases through emissions from land-use change," 1238.
70 Interview, European Commission official, June 2010. This interviewee recounted that "towards the end of 2007, we even had Tim Searchinger over here to give presentations to everyone in the Commission".
71 European Commission, *Annex to the Impact Assessment*, 144–145.
72 Adam, "Kelly orders biofuels review."
73 Edwards *et al.*, *Biofuels in the European Context*, 22.
74 Croezen *et al.*, *Biofuels*; and Edwards *et al.*, *Indirect Land Use Change from Increased Biofuels Demand*.
75 See Palmer, *Science and Politics in European Energy and Environmental Policy*.
76 European Commission, "Directive 2009/28/EC," 40.
77 Interview, expert scientist, April 2010.
78 Interview, environmental campaign group representative, July 2010.
79 Interview, European Environment Agency representative, July 2010.
80 Renewable Fuels Agency, *The Gallagher Review*; and Palmer, "Stopping the unstoppable."
81 European Commission, *Annex to the Impact Assessment*, 144–145.
82 Searchinger *et al.*, "Does the world have low-carbon bioenergy potential from the dedicated use of land?" 439.

83 A more detailed overview of efforts to develop "advanced" forms of bioenergy is provided in Chapter 4.
84 Searchinger *et al.*, "Do biofuel policies seek to cut emissions by cutting food?" 1421.
85 Searchinger *et al.*, "Does the world have low-carbon bioenergy potential from the dedicated use of land?" 439.
86 Kuchler, "Sweet dreams (are made of cellulose)," 431.
87 Interview, chemical engineering expert, June 2009.
88 European Commission, "Directive 2009/28/EC," 24 (emphasis added). The Directive includes a specific carbon accounting provision intended to incentivise the production of biofuels on what it terms "restored degraded land", defined as land which "for a significant period of time, has either been significantly salinated or presented significantly low organic matter content and has been severely eroded" (54).
89 Searchinger *et al.*, "Does the world have low-carbon bioenergy potential from the dedicated use of land?" 440; and Gibbs and Salmon, "Mapping the world's degraded lands." For comparison, world total agricultural land in 2022 was estimated by the Food and Agricultural Organization of the United Nations (FAO) at just under 4.8 billion hectares (FAO, *Land Statistics 2001–2022*).
90 Baka, "Making space for energy"; and Goldstein, "*Terra economica.*"
91 Searchinger *et al.*, "Does the world have low-carbon bioenergy potential from the dedicated use of land?" 440.
92 See Paprocki and McCarthy, "The agrarian question of climate change," 701.
93 Dauvergne and Neville, "Forests, food, and fuel in the tropics," 655.
94 Vermeulen and Cotula, "Over the heads of local people"; see also Neville, *Fueling Resistance.*
95 Burley and Griffiths, "*Jatropha*: Wonder crop?"
96 Baka and Bailis, "Wasteland energy-scapes"; and Baka, "Making space for energy."
97 Levidow, "EU criteria for sustainable biofuels," 211.
98 Due to the lower energy content of biofuels, any given barrel of bioethanol or biodiesel is equivalent, respectively, to only circa 0.6 or 0.85 barrels of crude oil (Energy Institute, *Statistical Review of World Energy 2024*, 71).
99 Bonneuil and Fressoz, *The Shock of the Anthropocene*, 101 (emphasis in original).
100 Simpson and Szeman, "Impasse time," 81; see also Fressoz, *More and More and More.*
101 Mayr *et al.*, "Palm oil, the RED II and WTO law."
102 See Cotula *et al.*, *Land Grab or Development Opportunity?* In a recent review of scholarship on global land deals, Wendy Wolford and colleagues position biofuels as integral to "ongoing green grabbing" into the 2020s (Wolford *et al.*, "Global land deals," 4).
103 Borras *et al.*, "The politics of biofuels, land and agrarian change," 583; see also Matondi *et al.*, *Biofuels, land grabbing and food security in Africa.*
104 Li, "What is land?" 592; see also Bridge, "Resource triumphalism."

105 See for example Carmody, *The New Scramble for Africa*; and Kugelman and Levenstein, *The Global Farms Race*.
106 Paprocki and McCarthy, "The agrarian question of climate change," 700; see also Huber and McCarthy, "Beyond the subterranean energy regime?"
107 Haiven, *Palm Oil*, 112.
108 Raman and Mohr, "Biofuels and the role of space in sustainable innovation journeys," 224; and Mol, "Boundless biofuels?" 310.
109 Juskus, "Sacrifice zones," 16.
110 Juskus, "Sacrifice zones," 16.

3

Growing energy on trees? Reinventing working forests as carbon conveyors

The pine trees seemed to go on forever, sweeping by in endless rows.

—William Boyd, *The Slain Wood*, 2015

On 13 March 2020, *Zheng Zhi*, a 220-metre bulk carrier vessel sailing under the flag of Panama, arrived at the British port of Immingham, situated on the Humber Estuary, with a cargo of more than 63,000 tonnes of combustible fuel.[1] Unlike most of the many thousands of other large cargo vessels that deliver fuel to the UK each year, however, this one was not laden with crude oil or liquified natural gas. The ship's cargo, in fact, did not comprise fossil fuels of any kind but rather a biomass fuel, in the form of compressed wood pellets. Moreover, these pellets had been derived initially not from British or even European trees, but from the extensive forests of the Southern US. Their voyage had begun several weeks earlier—and more than 4,500 miles away—at the most northerly deep-water port on the Mississippi River in Baton Rouge, Louisiana. Despite the apparent arduousness of the trip, ships like *Zheng Zhi* have been in increasingly good company in bringing wood pellets across the Atlantic in recent years.[2] Once safely docked, whether at Immingham or elsewhere, these wood pellets can finally embark upon the last leg of their long journey, being loaded onto custom-built freight wagons bound in most cases for Drax Power Station in Yorkshire, where they are then burned to generate electricity for the UK's national grid.

In the early months of 2020, the UK was embarking upon its longest continuous spell of coal-free electricity generation since 1882—a spell that would eventually last over two months.[3] Few commentators, however, credited this achievement to growing imports of foreign wood pellets. Instead, the anointed heroes in the story of the UK's rapid phase-out of coal—responsible for around a quarter of all the country's electricity in 2016, but only 1.8 per cent four years later—were more often expanding arrays of solar panels and wind turbines, whether onshore or offshore.[4] Yet with nearly 5 million tonnes of wood pellets being imported to the UK from the

US in 2023 alone,[5] and the combustion of such pellets at Drax being responsible by itself for around 8 per cent of the UK's overall renewable electricity supply,[6] forest-based bioenergy has clearly been indispensable to the realisation of a coal-free British energy system. So why are so few people talking about these imported wood pellets? And, perhaps more importantly, how has the UK come to rely so critically for the replacement of coal—and hence for meeting its own net zero climate change targets—not just upon solar panels and wind turbines on its own soil, but also upon the productivity of forests growing thousands of miles away in the US South?

This chapter will examine the rapid emergence of a burgeoning wood pellet manufacturing industry in the US South, focusing in particular on its consequences for understandings of the role that bioenergy, and indeed forests more generally, should play in tackling climate change over the coming decades. The production of wood pellets from trees has increasingly been incentivised by national governments as a means—officially at least—of bringing into being more sustainable and climate-friendly energy systems not just in the UK,[7] but also elsewhere in Europe and indeed further afield in countries like Japan and South Korea.[8] While compressed wood pellets have been produced in large volumes across North America, Europe and beyond since 2010, however, the forests of the US South region have proven particularly attractive to energy providers seeking to procure this ostensibly renewable alternative to coal.[9] The chapter therefore begins by asking what is so special about the forests of this region that it can make sense to ship wood pellets across the Atlantic in such vast quantities, and by scrutinising the case for viewing such arrangements as beneficial for climate change in greater detail.

The emergence of new international supply chains for combustible biomass fuels does not just have profound implications for the geography and geopolitics of contemporary energy systems, however. In linking energy consumption directly to the production of a more stable global climate, these supply chains also open up pathways for the realisation, from quite longstanding forms of industrial forestry, of new forms of economic value. Moreover, these new value production pathways are unique within the energy sector for their dependence, at least in part, upon the vital capacities of trees to photosynthesise, grow and sequester carbon dioxide. Consequently, later sections of the chapter will examine the specific arrangements of human labour and vegetal growth—or indeed, vegetal labour—upon which efforts to maximise value production in the forest-based bioenergy sector are mounted.[10] As more countries seek to replace coal with biomass as an electricity-generating fuel, interrogating the forms of labour which animate working forests in the US South (and indeed elsewhere) helps to illuminate the shifting ecological basis for value production in the global energy sector today. More fundamentally, however, a focus on labour can also help to reveal how these forests are increasingly viewed as dynamic, living embodiments of particular understandings of value itself, serving in effect to naturalise energy cultures predicated upon the pursuit of efficiency and productivity not for higher ends, but simply for their own sakes.

The US South: wood basket of the world?

For geographer David Demeritt, settler American identity was shaped principally "not by force, or ethnicity, or religion, so much as by [a] common experience of the forest frontier".[11] Today, many of America's forests are among the most economically productive in the world. Few would dispute that the "common experience" of modern American forestry was initially marked, however, by considerable recklessness. Particularly from the second half of the nineteenth century, vast swathes of the nation's extensive old-growth forests were rapidly exhausted at the hands of a "wave of essentially extractive forestry", impacting first the forests of the Northeast before moving quickly through the Great Lakes, the South and Southeast and eventually even the Pacific Northwest.[12] By the mid-twentieth century, much of the forest that would have greeted initial European settlers of colonial America had already been converted, by a rapacious lumber industry, into "a landscape of devastation".[13] In subsequent decades, where forest has been replaced, it has not typically been with diverse tree species intended to mature to old age and fulfil diverse ecological functions, but rather with timberlands whose purposes are circumscribed in much narrower, starkly economic terms. Accordingly, much of the more than 3 million square kilometres of forested land found in the US today is referred to colloquially as "working forest", denoting its management expressly for the production of wood-based commodities.

Even accounting for the recent financial woes of the housebuilding and construction industries,[14] as well as a more gradual and ongoing decline in demand for paper, the US forest products industry of the twenty-first century remains disproportionately productive. Indeed, despite possessing only around 8 per cent of the world's forests, the US accounts for almost 20 per cent of present-day industrial roundwood production and close to 30 per cent of global forest products.[15] While they may not contain the largest or most economically valuable trees (those accolades belonging unequivocally to the colossal Douglas firs of the Pacific Northwest), the forests of the US South region specifically account for more than their fair share of this overall productivity. Approximately one million square kilometres, or one third of all the nation's forest, as well as 70 per cent of its deliberately planted timberland, are concentrated into the thirteen states whose collective footprint runs from Texas to Florida along the Gulf Coast, and is bounded to the north by Oklahoma, Arkansas, Kentucky and Virginia.[16] In contrast to the Pacific coast region, where 68 per cent of all forest is under public management, well over 80 per cent of Southern US forests are privately owned. Family-owned plots, moreover, while often comparatively small on their own terms, account for twice as much forest as typically more extensive, commercially owned stands.[17] Despite their heavily fragmented nature, however, the forests of the South witnessed almost 200 million cubic metres of timber removals in 2013, or 58 per cent of the national total.[18] This, then, is not only the most heavily forested region in the US, but also the most intensively exploited.[19] In short, the South is the undisputed "wood basket of the Nation", and for some the "wood basket of the world".[20]

Against this backdrop, it is perhaps unsurprising that bioenergy industry actors have taken a particular interest in the forests of the US South in recent years. After all, the region is already well accustomed to the presence of diverse forest product industries. For many decades, communities here have relied significantly for employment not only on mills set up to process large-diameter saw-logs into furniture or construction materials, but also on paper and pulp mills making use of smaller-diameter trees, wood chippings and other residues. Importantly, this is not to say that power relationships between government, businesses and communities in the region are equitable or devoid of rancour; much of the South is "steeped in a history of economic dependence on forest products, and on exploitative systems of labour".[21] Since the mid-1950s, however, the region's forests have exhibited marked productivity increases in response to rising demand for all kinds of wood-based products. Not only has the overall area of unreserved timberland in the South remained relatively stable throughout this period, but the total repository of wood growing on those lands has steadily increased, in some estimates more than doubling by 2015 against mid-twentieth-century levels.[22] Much of this increase can be attributed to the long-term success of industrial tree improvement programmes, which, since the 1950s, have used selective breeding techniques to gradually develop more vigorous genotypes of commercially important species.[23] Loblolly pine (*Pinus taeda*) is exemplary in this respect; final-harvest tree volumes for this species are estimated to have increased by 20–35 per cent when compared to the "unimproved seedlings" of a few decades ago.[24] While loblolly is by no means the only species of commercial use cultivated in Southern forests (being accompanied in many places by hardwoods like oak, sweetgum, hickory and cypress), it is by far the most prevalent; there are thought to be around 22 billion individual loblolly pines growing in the South today, accounting for 20 per cent of the region's entire above-ground biomass.[25]

Taken in the round, these figures underline the status of forests of the US South as archetypal examples of landscapes historically shaped by efforts to "directly augment natural processes and use them as strategies for increasing productivity".[26] Particularly when set against the aforementioned declines in conventional paper production, this story of a regional forest resource whose productive potential is steadily increasing appears to offer fertile ground for protagonists of wood pellet manufacturing as inherently beneficial for the US South. Indeed, manufacturers have not been reticent to assert a need for economic "revitalization"—in short, for jobs—when underscoring the benefits that wood pellet plants allegedly bring to local communities.[27] Yet the fact remains that the growth of this industry, driven initially by policies implemented in the UK and Europe, has been not only dramatic but also deeply controversial. As recently as the late 2000s, there were no major wood pellet manufacturing facilities to speak of in the South. By 2023, no fewer than twenty-eight industrial-scale pellet mills enabled the region to export just over 9.5 million tonnes of pellets, with the three largest beneficiaries being the UK, Japan and the Netherlands.[28] As exports have grown, concerns have increasingly been raised not only about the alleged sourcing of feedstocks from unmanaged, ecologically sensitive areas, such as the bottomland hardwood forests of the South's coastal plains,[29] but also about the effects, suffered especially by "rural, low-wealth communities of color", of various forms of pollution associated with the manufacturing of wood pellets themselves.[30] Indeed, according to one study, wood pellet plants are 50 per cent more likely to be sited in recognised "environmental justice communities"—defined as counties where poverty levels exceed the state median and at least 25 per cent of the population is non-white—than elsewhere.[31]

Even if these important social concerns are overlooked, however, the argument that some wood pellet manufacturing facilities have provided welcome jobs to replace those lost to declines in paper and pulp production nonetheless serves to conceal critical differences between the bioenergy sector and other forest-based industries historically active in the region. To be sure, wood pellet manufacturing and paper and pulp production require very similar kinds of biomass feedstocks. Yet the pathways through which the bioenergy industry generates value from those feedstocks are more complex than those to be found in extant industries. For wood pellet manufacturers, forests are no longer viewed exclusively as repositories of material for wood-based commodity production, effectively serving as living warehouses. Instead, they constitute dynamic ecological systems whose improved management might just serve to turn energy (and specifically electricity) consumption—at least in foreign countries, if not in the US itself—into a climate change mitigation tool. Wood pellet manufacturing, in other words, amounts to a new kind of resource-making, one concerned not just with tangible fuel production, but with generating value also from its claimed links to a wider process of climate repair. Accompanying these shifts, perhaps inevitably, are accusations that all is not as rosy as it seems.

Burned? Counting the costs of wood pellet manufacturing

From a legal perspective, the argument that wood pellet combustion is beneficial for the global climate system does not need to be made; it is already enshrined in international carbon accounting frameworks agreed under the auspices of the UNFCCC—and in official UK and EU policies—through the straightforward categorisation of all biomass fuel as carbon neutral.[32] As discussed in Chapter 1, this approach effectively discounts the carbon dioxide emissions generated upon burning biomass, despite the fact that those emissions are in fact more intensive, per unit of energy generated, than those associated with burning coal.[33] Instead, emissions linked to biofuel combustion are counted within the land-use sectors of countries where biomass feedstocks are initially grown.[34] Nonetheless, carbon still constitutes "a troublemaker" in efforts to smoothly expand the modern bioenergy sector today.[35] For critics of forest-based bioenergy specifically, this amounts to a regulatory sleight of hand hinging on the "false assumption" that carbon emissions arising from biomass burning will always be "quickly offset by forest growth somewhere else".[36] Accordingly, scientists have criticised prevailing bioenergy carbon accounting regimes for overlooking a "carbon debt" which could, where fuels are derived from forests, take several decades to eradicate through the growth of replacement trees (assuming those replacements are planted at all).[37] As one prominent letter, signed by over eight hundred scientists and delivered to the European Parliament in January 2018, put it: "At a critical moment when countries need to be 'buying time' against climate change, this approach amounts to 'selling' the world's limited time to combat it".[38]

When it comes to the US South's burgeoning wood pellet manufacturing industry, these concerns have been significantly amplified through effective environmental campaigning on the ground. Images of swathes of clear-cut forest, along with accounts appearing to show wood arriving at pellet mills from logging sites in mature forests, often with high ecological significance, have registered with wider audiences in a manner that no critique of carbon accounting methodologies alone could ever have hoped to achieve. Indeed, campaigns such as "Our Forests Aren't Fuel"—jointly coordinated by the Natural Resources Defense Council and Dogwood Alliance—have been assiduous in portraying carbon accounting not as an esoteric legal matter, but as a political tool marshalled to subordinate communities and environments of the US South to the strategic needs of governments, energy companies and ultimately electricity consumers in the UK, Europe and elsewhere.[39] From the perspective of these campaigns, the South should be understood as nothing less than "a sacrifice zone for the forestry industry, which is now parading as a clean energy industry in the form of biomass".[40] Detailed concerns raised by these campaigns relate not only to increased forest degradation and impacts on biodiversity, but also to the bombardment of the disproportionately low-income, non-white communities living near pellet plants themselves with noise, dust and air pollution—the latter often including hazardous compounds such as methanol, acrolein and formaldehyde.[41] Pellet plants have also proven vulnerable—perhaps unsurprisingly—to the outbreak of fires, which not only generate large pulses of air pollution but also can prove extremely hard to extinguish.[42] That such afflictions have arisen from a new industry subsidised not only indirectly by governments abroad, but also often directly by US federal and state-level grants, has only heightened a widespread sense of "disillusionment and distrust" with policymakers and businesses alike in the region.[43] Indeed, as Sarah Hitchner and colleagues have observed, and despite claims that pellet manufacturing has created valuable new jobs, many rural residents in the South ultimately feel "betrayed" by government support for the industry.[44]

Faced with these diverse complaints, advocates of forest-based bioenergy have sought to protect the reputation of wood pellet manufacturing in the US South by insisting on particular approaches to counting its costs. On one level, these approaches rely on keeping the social costs of pellet manufacturing out of sight, at least as far as possible. Thus, while local community testimonies have been influential in shaping conversations in the US South itself,[45] from the standpoint of debates about forest-based bioenergy taking place in the UK and Europe, accounts of these forms of "slow violence" have struggled to gain traction, suffering from what Thom Davies has called "a politics of indifference".[46] Indeed, however "arresting" they might be to those who hear them firsthand, stories about the lived experiences of pollution around pellet plants have been routinely trivialised and downplayed in the more rarefied scientific and policy debates about bioenergy taking place in London and Brussels; simply put, for those living a safe distance away, these are accounts that, all too often, "do not count".[47] Moving away from on-the-ground social impacts of pellet manufacturing and towards its implications for global climate change, meanwhile, forest-based bioenergy's protagonists have also sought to insist on particular ways of counting—this time of GHG emissions. More specifically, industry protagonists have increasingly allowed themselves, ironically in common with many environmental campaign groups, to acknowledge deficiencies in official carbon accounting methodologies for biomass fuels, instead adopting a more fine-grained focus on the metabolic realities of forest growth on the ground. What really matters, from this perspective, is not the standalone emissions "footprint" of wood pellet manufacturing per se. Rather, attention is turned to bioenergy's impacts on broader market dynamics, wherein a complex ecosystem of forest product industries *collectively* shapes the health and productivity of the region's forest resources.

Certainly, it is true that no timberland is ever managed to produce just one type of wood-based commodity. In working forest stands dominated by loblolly pine, for example, landowners typically derive income not only from the sale of large-diameter saw-logs at the end of the growing cycle (with final harvests taking place around 25–35 years after initial planting), but also from one or more intermediate "thinning" operations, in which a proportion of smaller-diameter trees are removed—for loblolly pine, the first such operation would typically occur after 12–15 years.[48] By the time a stand is ready for thinning, however, the pines themselves will almost always have been joined—unbidden—by numerous hardwood trees whose seedstock is ubiquitous in soils across the region, and whose growth is therefore almost impossible to repress without herbicides.[49] A typical thinning operation therefore yields not only small-diameter roundwood (the main trunks of 12–15 year old loblolly pines), but also wood chips derived from smaller hardwood trees, as well as tree limbs and tops that would otherwise be of little commercial value. When the time comes for a final harvest, meanwhile, things are even more complex, with diverse classes of smaller- and larger-diameter roundwood usually being available, some of much greater commercial value than others, as well as significant additional woody biomass comprising branches and tops that can be turned into chippings. In view of this complexity, the economic success of logging operations in the US South relies heavily on the "craft of decision-making" of an on-the-ground workforce seeking to optimally allocate diverse woody feedstocks into three, four or even—for very complex, mixed-age stands—as many as ten "separations", each bound for a distinct end market (see Figure 3.1).[50] Attempts to calculate the precise climate change impacts of any single forest product industry, meanwhile—whether linked to the bioenergy sector or not—quickly run up against uncertainties around the question of who exactly is accountable for forest removals in the first place.

By foregrounding these dynamic interactions, industry protagonists aim to depict wood pellet manufacturing as sustainable not because all biomass fuels are automatically carbon neutral, but rather because the bioenergy sector is inherently subordinate to other forest product industries active in the US South. As one wood pellet mill operative in North Carolina put it: "[A]nything larger than 18 inches in diameter has no business being here;

Figure 3.1 An active logging deck in Duplin County, North Carolina, May 2019. Several "separations" are discernible, including (left) a pile of small-diameter roundwood stems, (centre) a consignment of larger-diameter roundwood stems loaded onto a truck and (right) smaller, leafier tree limbs being processed into wood chippings.
Source: Author's own photograph.

it was growing to be something other than a pellet".[51] For another forest product industry representative, more bluntly, any case for wood pellet manufacturing being good for climate change would simply "fall apart if wood was being grown specifically for bioenergy".[52] By adopting this stance, wood pellet advocates have importantly not sought to deny that their industry does make use of whole trees. In a public interview given in early 2018, for example, the secretary general of the Confederation of European Forest Owners was at pains to insist that "when you get down to lower quality wood, bioenergy can be a valid option if the wood cannot be used by a sawmill, even if it looks like a normal, big tree trunk".[53] Crucially, by adopting this stance, representatives of the wood pellet industry have also enabled themselves to admit that "not all biomass should automatically be categorised as carbon neutral".[54] Environmental campaign groups and other critics of prevailing carbon accounting frameworks for biomass fuels have thus found themselves joined, in their opposition to these approaches, by the very industry actors whose commercial activities that opposition is intended to undermine. Yet defenders of forest-based bioenergy production have sought to do much more than merely take the sting out of prevailing controversies about the industry's carbon footprint. As the next section will show, some have also mounted a more bullish case for viewing bioenergy production as key to actively *improving* the productivity—and indeed the environmental performance—of Southern working forests as a whole.

Silvicultural synergies: optimising forest work

On the surface, the idea that an increased use of wood might not deplete forests, but augment them, seems anathema to basic common sense. Historical data, however, do lend credence to this view. According to one prominent US Forest Service report, between 1970 and 2010, even as the South experienced significant population growth of 88 per cent, the volume of timber products derived from the region's forests more than doubled, while total forest biomass increased by more than 50 per cent, all within a slightly reduced area of forested land.[55] Reductions in forest extent, meanwhile, have been shown to be driven significantly by factors unrelated to demand for wood products, including notably urbanisation; according to one estimate, close to half of all forest land lost to other uses between 1982 and 2012 was converted for development purposes.[56] In light of these trends, it is an oft-repeated mantra in the South's forestry sector today, and not just among representatives of the bioenergy industry, that the best protection for the region's forests can be found not in regulating wood-based commodity markets, but rather in strengthening them. Flourishing markets for wood, from this perspective, are not a death knell for forests, but rather the best available means of ensuring that forested land is replanted after it has been harvested, rather than being turned over for other uses.

From the perspective of Drax, one of the most prominent advocates and indeed beneficiaries of wood pellet manufacturing in the US South, the lessons emerging from these historical data can be expressed very simply: "The more wood we use, the more forests grow".[57] Such pithy refrains, however, belie much complexity in the mechanisms through which productivity increases have actually been achieved in forests of the US South in the past. While the aforementioned industrial tree improvement programmes have been pivotal in producing more vigorous trees, for example, this is far from the only way in which forestry capital has engaged, particularly since the mid-twentieth century, in "ongoing intervention in and alteration of tree growth processes" in the region.[58] No less important, though certainly less glamorous, has been the gradual refinement, over many decades, of routine silvicultural practices which seek to boost forest productivity by operating at the landscape level rather than the genetic level, targeting interactions between the various trees growing in a stand. Of particular significance in this regard has been the longstanding practice of "thinning". If enacted well, this apparently mundane intervention promises not only to avail landowners of intermediate income from the sale of small-diameter roundwood and wood chippings, but also to enhance the size and quality of those trees left to continue growing into high-value saw-logs. Intriguingly, however, a defining feature of arguments in favour of forest-based bioenergy production in the US South since 2010 has been the reinvention of this practice as central not just to improving the productivity of the region's forests, but also to enhancing their capacity to contribute to climate change mitigation.

The primary objective of thinning is straightforward—to remove a moderate proportion of trees from a stand in order to pre-empt competition for light and space, thereby enabling the remaining trees to grow unimpeded into the best possible quality saw-logs.[59] For loblolly pine trees, as mentioned, thinning operations are typically carried out around 12–15 years after initial planting has taken place.[60] Several factors militate against simply avoiding the need to thin by planting fewer trees in the stand to begin with. Firstly, the presence of a significant seedstock of native hardwood trees in local soils means that juvenile loblolly pines typically face considerable competition from other species in the first few years after planting; denser planting arrangements—around 1,200 trees per hectare is not uncommon—help to reduce the inroads that oak, maple and other hardwoods can make into the stand in the crucial initial stages of growth.[61] Secondly, even in the absence of unwanted competition from hardwood trees, pines which are planted too far apart from one another will be more likely to direct some of their energy into expanding outwards, rather than growing upwards, thereby compromising the eventual length and quality of the saw-logs which can be derived from them at the point of final harvest. No less important than either of these two considerations, however, has been a historically buoyant demand in the US South—particularly owing to the success of the paper and pulpwood sectors—for precisely the kinds of small-diameter roundwood and wood chippings which thinning operations yield. The ability of wood pellet manufacturing to compensate for recent declines in demand for paper, accordingly, is often depicted by industry protagonists as a welcome source of stability for the region's forestry sector, since it acts to perpetuate a tried and tested formula for maximising value from the South's working forests, based around dense overplanting, vigorous initial growth, and astute, well-timed thinning. Whereas the initial overplanting of stands is a relatively straightforward procedure, the determinants of well-executed thinning are more complex, and worthy of further scrutiny.

To begin with, when loblolly pine trees begin to encroach upon their neighbours, deciding which individuals to remove and which to leave in place entails paying close attention to the unique patterns of arboreal growth and interaction which have shaped the character of the stand as a whole. Particularly weak or poorly formed trees are usually not hard to identify, of course, and must be removed as a matter of priority to enable "growth to be put on the better trees in the stand".[62] Eradicating weaker individuals, however, almost always accompanies removing plenty of perfectly healthy, straight and vigorously growing trees from the stand as well, not least in order to clear space for heavy-duty logging machinery to do its work (see Figure 3.2). As one wood pellet industry representative put it in an interview, "[Y]ou lose some good trees, but you also get rid of all the weaker ones".[63] Deciding how many healthy trees to remove, and indeed which healthy trees to remove from a stand, however, is never an exact science but rather a judgement call. As a result, considerable emphasis is placed upon hiring forestry professionals who can be trusted to undertake thinning astutely; likewise, a logger's reputation can be made or lost on the basis of their ability to correctly foresee how thinning will reshape interactions between trees left in place over the years, or even decades, following the intervention itself. While the optimal "target stocking density" for a stand of thinned loblolly pine trees can be expressed in relatively precise terms (usually as a number of tree stems to be left in place per hectare), it can still take many years for foresters to "get a very good feel" for how to achieve that density in a manner that maximises future growth rates while

Figure 3.2 A recently thinned stand of loblolly pine trees in North Carolina, May 2019. In order to achieve optimal spacing for trees to grow into high-value sawlogs, a proportion of perfectly vigorous and healthy pine trees have been removed from this intensively managed stand.
Source: Author's own photograph.

also minimising the potential for timber quality to be compromised by deformities in tree shape.[64] In short, thinning is not an operation that can be distilled down to codified, step-by-step instructions; instead, it is a form of labour which relies fundamentally on "affective ways of knowing nature".[65]

No less crucial to the success of thinning operations, moreover, is judging whether the time is right to thin in the first place. Pinpointing when it is "silviculturally sensible to intervene",[66] as one interviewee put it, can significantly impact how much value is derived from a stand of loblolly pine trees in the course of a full growing cycle. Commit to thinning too soon and not only will the quantity of small-diameter roundwood derived from the operation be suboptimal, but remaining trees will also be more likely to "broaden out" rather than growing as tall as possible.[67] Postpone thinning for too long, by contrast, and the consequences can be just as costly. Having been forced into close proximity with their neighbours, trees in these stands will typically possess smaller "live crowns"—the uppermost part of the tree, where photosynthesising needle leaves can be found (see Figure 3.3). Consequently, they will be less able to capitalise upon newly available light and space following thinning and, as one forestry professional put it, "[S]tand growth rates will never recover".[68] While the average

Figure 3.3 A recently thinned stand of loblolly pines in Mississippi, April 2018. Compared with the stand pictured in Figure 3.2, many of the live crowns here are sparse, and some are very small relative to the overall height of the trees. This stand is not intensively managed and has been thinned later than would be optimal if the landowner were seeking to maximise subsequent tree growth rates.
Source: Author's own photograph.

diameter of tree stems and, indeed, their total basal area are both important indicators of a stand's readiness for thinning, perhaps the single most important variable to which foresters pay attention when taking this decision is the "live crown ratio", measured typically as the proportion of a tree's total height playing host to live needle leaves at any given moment. That such emphasis is placed on optimising this particular characteristic reveals that even if loblolly pines in working forests have only recently begun to be used as bioenergy feedstocks, they have long been regarded in expressly energetic terms. Indeed, that thinning is ultimately concerned with maximising the efficiency of the energy conversions performed by trees was underscored in an interview by one wood pellet industry representative, for whom the issue of live crown size ultimately boils down to whether a tree has "a big enough engine to have the best possibility to achieve full log size".[69]

Although it has typically been regarded as a craft skill, then, thinning is clearly also an intervention into forest growth dynamics that is no less stubbornly concerned with maximising profits than the practice of selectively breeding trees to promote more vigorous genotypes. Indeed, from the perspective of critical geographical scholarship, thinning can be interpreted as a key component of wider strategies aiming to make trees "work harder, faster and better" in industrial forestry, thereby achieving what William Boyd and colleagues have called the "real subsumption of nature".[70] Yet the challenges entailed in executing thinning to a high standard reveal that humans can never simply plant, manage and harvest trees according to whim. Rather, the forms of human labour which are brought to bear on any given stand of trees will always unavoidably be shaped by the performances—expressed in terms of non-uniform, indeterminate growing patterns and dynamics—of the trees themselves. Moreover, and just as intriguingly, while thinning has been pivotal to the steadily increasing overall productivity of Southern working forests in the second half of the twentieth century, the region's incumbent forest product industries have typically placed little emphasis upon it, preferring instead to defend their practices through reference to the overarching principle of sustained yield.[71] By contrast, and as the next section will set out, wood pellet manufacturers have been more assertive in positioning thinning as a means not merely of ensuring that forest yields are sustained over time, but of actively increasing the efficiency of the energetic conversions that trees perform while growing. In so doing, these actors have evinced a broader view of industrial forestry in which humans must find ways to work *with* trees—in a synergistic fashion—and not merely seek to work upon them, as if they were an infinitely pliable material substrate. Trees, in short, are coming to be understood as dynamic players whose vital capacities actively shape the remit of forest-based economies, and not merely "as 'ready-made' objects" to be used by capital in a more or less economic manner.[72] Those vital capacities, moreover, extend not just to trees' abilities to grow, but also to their abilities to influence their surroundings—and indeed the wider environment—as they do so.

Inventing the forest carbon conveyor

In many ways, the reliance of wood pellet manufacturing upon the kinds of feedstocks typically yielded during thinning operations might appear to downplay the difference that this new industry makes to historically dominant forestry management practices in the US South. Yet, in emphasising the benefits that thinning brings to the productivity of working forests as a whole, bioenergy protagonists have also laid the foundations for a quite different way of thinking about the role that trees should play in contributing to not just sustainable commodity production, but also climate change mitigation. At the crux of this perspective is the question of what it means, precisely, for forests to grow. For while efforts to evaluate the impacts of bioenergy production in the US South (and elsewhere) frequently focus on the growth (or loss) of forests themselves, where what is measured are changes in the total area of forested land, perfectly valid questions can also be asked about the industry's impacts upon growth rates within forests, where what is measured is instead the intensity of the forest's overall metabolic activity. To defend their claim that "The more wood we use, the more forests grow", therefore, prominent bioenergy industry actors do not necessarily need forests themselves to become more extensive. Instead, they point to alleged accelerations in overall tree growth rates taking place across the region, as more and more landowners ostensibly seek to capitalise on rising demand for wood by engaging in more intensive and active management of their stands. Such arguments, of course, significantly downplay the wide range of factors which can shape planting and harvesting decisions made by forest owners across the region, especially in family-owned contexts where there is often significant attachment to "values such as beauty or scenery, nature protection, wildlife habitat, hunting, recreation, family legacy, and investment".[73] Backgrounding this complexity, however, enables bioenergy advocates to emphasise the difference that a rising demand for wood pellets should make—at least in theory—to the internal metabolic dynamics of working forests already under intensive management.

From this perspective, it matters critically that younger loblolly pine trees—partly owing to the historical success of industrial tree improvement programmes—tend to grow much more vigorously than their older counterparts, not only achieving greater proportional volume increases but also sequestering more carbon dioxide relative to their size, in any given time period.[74] As major UK wood pellet importers Drax have explained in their own exposition of working forest management: "[O]lder trees will have more carbon stored (after a 'childhood' spent absorbing it), but if these are not harvested ... their carbon absorption plateaus".[75] What is at stake here is not simply a claim about the environmental sustainability of wood pellet manufacturing, measured on its own terms, so much as the establishment of an entirely new moral economy of tree life in working forests—one which actively discriminates among trees on the basis of the rate of the work they can be made to perform.[76] Given their tendency not just to grow more rapidly but also to sequester carbon dioxide more efficiently, younger, smaller trees emerge clearly within this moral economy as more virtuous than their older, larger (but nonetheless slower-growing) equivalents.[77] The broader implications for efforts to maximise the ability of working forests to contribute to climate change mitigation are clear; while older trees may hold large amounts of carbon dioxide in situ, they also occupy considerable space which could alternatively play host to larger numbers of younger, more dynamic and vigorously growing trees. From this vantage point—and particularly given that working forests are already being managed to produce a range of wood-based commodities in any case—it makes sense to enhance the ability of these stands not strictly to store large amounts of carbon, but rather to actively sequester carbon at the fastest possible rate. Tree size and stability, in short, are far less important to the health of the working forest than arboreal dynamism and productivity.

In seeking to counteract claims that forest-based bioenergy leads inevitably to a "carbon debt", and that it may in some cases be worse for climate change than simply continuing to burn coal, protagonists of wood pellet manufacturing have thus effectively sought to appropriate, and subtly reinvent, the idea of the working forest itself. Previously a somewhat unassuming label denoting any forest stand being managed deliberately for commodity production, the idea has been carefully recast by bioenergy interests as a silvicultural paradigm capable not simply of sustaining the intensive production of wood-based commodities, but rather of actively enhancing the rate at which carbon dioxide is brought out of the atmosphere and transmitted *through* the forest into a diverse range of tangible commodity forms. While the precise silvicultural practices undergirding working forest management—dense overplanting of stands, vigorous initial growth, and astute, well-timed thinning operations—are not new to the US South, no industrial actor has previously sought to establish a causal link between those practices and wider benefits for the global climate system. To grasp the full benefits of bioenergy production, according to this outlook, one must recognise the industry's growth as a driver of increasingly synergistic silvicultural practices and working arrangements throughout the region's forests, bringing productivity increases to diverse forms of commodity production simultaneously. As one bioenergy industry representative explains, even if wood pellet manufacturing frequently constitutes a "natural side-stream from high-value timber production", it also acts to "really contribute beyond its direct footprint ... bringing benefits throughout the forestry sector".[78] Moving to claim responsibility for the enhanced production of *all* forest products in this way is particularly significant from a climate change perspective, because carbon dioxide tends to remain locked up in most wood-based commodities—particularly those destined for use in the furniture and construction industries—for a much longer time than it does in wood pellets themselves.[79]

The thrust of these arguments, of course, pays little heed to alternative strands of forestry science, such as those which emphasise the often-immense ecological value of mature forests, or the significant contributions that can be made to climate change mitigation and wider ecosystem services by unmanaged forests playing host to a more diverse range of tree species, rather than commercial plantations.[80] Recent research has, for example, pointed to wide discrepancies in the carbon sequestration capacities of planted forests managed intensively for commodity production compared to naturally regenerating stands, indicating that the arguments being advanced by wood pellet manufacturing's protagonists are highly context specific.[81] On a more fundamental level, meanwhile, a focus on maximising rates of carbon sequestration achieved by individual trees also overlooks prominent work in forest ecology highlighting the vital importance of soils, and especially of mycorrhizal fungi, in facilitating the storage and exchange of carbon within forest ecosystems over long time periods.[82] Even if we accept that an energetic lens is appropriate for assessing the overall health of a forest, moreover, not all forms of energy are strictly thermodynamic. In longstanding regimes of regenerative forestry management practiced in Wisconsin by members of the Menominee Tribe, for example, the energy of the forest constitutes "a set of social relations among all beings", with important cultural and spiritual dimensions which transcend any narrow concern with metabolism, work or productivity.[83] Nonetheless, at least in the specific context of working forests already being managed for commercial purposes, the central idea being advanced by bioenergy producers is that efforts to tackle climate change can be best served not strictly by increasing the amount of carbon stored in these stands, but rather by achieving the more rapid and efficient movement of carbon through them and on to other places (where that carbon could instead be stored in diverse commodity forms ranging from paper and cardboard boxes to roof beams and writing desks).

Ultimately, then, this is a view of working forests being deliberately managed to contribute to climate change mitigation not by operating as gradually accumulating carbon sinks, but rather as vigorous, high-throughput "carbon conveyors".[84] According to these arguments, were wood pellet manufacturing to cease, this would act not merely to reduce the available quantity of biomass fuel with which to replace coal in power stations in the UK, Europe and elsewhere—it would also compromise the ability of working forests in the US South to convey carbon dioxide out of the atmosphere and into other commodity forms as well. In this context, and despite being a longstanding silvicultural staple in the region, the enactment of well-timed thinning operations is effectively transformed—in the hands of the bioenergy sector—into a practice which serves to add value not just to those trees left in place to continue growing into high-quality saw-logs, but also to those trees removed and processed into wood pellets themselves. In other words, for those who subscribe to this standpoint, wood pellets can be said to embody value not just in the immediate physical sense, as combustible energy carriers, but also in a more symbolic sense, as signifiers of an increase in the overall work rate—and hence throughput—of the forest carbon conveyors from which they were derived.[85] The principal hallmark of a well-managed working forest hence comes to be redefined, no longer merely as the achievement of sustained yield, but rather, in the words of Drax, as the simultaneous presence of "a steady stream of CO_2-hungry young trees and a steady stream of renewable raw material".[86]

Bioenergy and forest futures: towards collaborative carbon metabolism?

Given that industrial forestry has sometimes been described in dismissive terms as "farming in slow motion",[87] it is perhaps unsurprising that the guiding imperative of almost all historical interventions into commercial working forests—at least in the US South region—has been the acceleration of forest metabolism. Nonetheless, given that the majority of timber has the potential to be transformed into a diverse array of different end products, and that significant market fluctuations can occur during the time required for a loblolly pine tree to reach full saw-log size, the relationships linking individual forest stands with particular industries—unlike those linking individual fields of crops to specific end markets in agriculture—are never stable or fixed in advance, but inherently provisional and open to future reconfiguration. Similarly open to revision, as a result, are understandings of the relationships that should link forest productivity in the abstract with specific economic or indeed wider societal objectives. And in this respect, unlike historically incumbent commercial actors in the forests of the US South, wood pellet manufacturers and bioenergy interests have sought to enrol these landscapes explicitly within efforts not just to produce sustainable wood-based commodities, but also to deliberately influence the composition of the global atmosphere. Their success in this endeavour, I have argued in this chapter, owes much to a "politics of indifference" in bioenergy-consuming countries about the "out of sight" but nonetheless significant social impacts of wood pellet manufacturing on the ground, borne disproportionately by low-income communities of colour in regions of the US South already heavily afflicted by environmental injustice.[88] But it has also been built on a subtle conceptual reinvention of working forests as more than simply living warehouses managed for tangible commodity production. More ambitiously, these forests are now to be viewed as dynamic, high-throughput carbon conveyors, acting in effect as photosynthetic engines serving to extract carbon dioxide from the atmosphere and move it on to other places as quickly and continuously as possible.

From the perspective of bioenergy interests, perhaps the most pragmatic implications of this account of working forests are those which relate to longstanding, thorny controversies over carbon accounting. By insisting that the climate impacts of wood pellet manufacturing cannot meaningfully be disentangled from those of the wider production systems which animate working forests as a whole, efforts to calculate the precise carbon footprint of any single forest product industry are rendered into a moot point. Somewhat more fundamentally, however, this view also has the effect of shifting the focus of debates about forest-based bioenergy away from the eminently fragile and contestable regulatory classifications of biomass fuels as carbon neutral, and instead towards the dynamic, metabolic processes taking place within working forests themselves. The ultimate proof of bioenergy's climate benefits will not be found, these arguments suggest, in any lofty lifecycle analysis of the sector's GHG emissions, but rather in the more tangible, and hence ostensibly verifiable, existence of trees growing more quickly and efficiently in forests than they were before. In adopting this position, protagonists of wood pellet manufacturing equate the value of trees themselves not strictly with the size or quality of the wood they comprise, and much less still with the contributions those trees might make to sustaining diverse ecological or indeed cultural lifeworlds. Rather, they focus narrowly on the rate at which trees photosynthesise, grow and sequester carbon dioxide. A hierarchy of preferences for trees is thereby established based explicitly upon their relative productivity; harder-working, more vigorous trees are regarded as more virtuous and respectable than lazier, relatively moribund ones.[89] This hierarchy, moreover, serves effectively to project a labour theory of value onto the vegetal world itself, rendering the performance of hard work—whether by humans or by trees—into an incontestable moral good.[90]

While this view of working forests as carbon conveyors has emerged from within the context of a geographically situated production system which is, in many ways, unique to the US South region, the fact that this system is playing a crucial role within international efforts to mitigate climate change is only likely to increase its impact within broader, ongoing debates about future forest governance. Certainly, the idea that trees can contribute best to environmental goals by working harder offers little resistance to the ethos of influential eco-modernist narratives, which—while rejecting conservative accounts of economic growth and environment protection as fundamentally contradictory—nonetheless reinforce those accounts' familiar faith in "the unquestioned value of work".[91] Yet it is also the case that trees in working forests of the US South are not being asked to perform just any old kind of work. Rather, in the bioenergy industry's view of these forests as carbon conveyors, there is evidence of a conscious disaggregation of trees' vital capacities on the one hand to photosynthesise and grow (thereby drawing carbon dioxide down from the atmosphere), and on the other hand to bring solar energy to rest in the form of wood (thereby storing carbon dioxide in situ). The ultimate goal of working forest management, at least when it is understood as contributing to climate change mitigation, is thus to enhance and optimise the rate at which forests extract carbon dioxide from the atmosphere and move it on elsewhere. The capacity of trees to hold on to that carbon dioxide as part of their own biomass over time, by contrast, is increasingly rendered obsolete by virtue of the ability of capital to produce (and generate demand for) wood-based commodities which can perform this function of their own accord (as for instance when construction materials derived from working forests lock carbon dioxide up in the built environment). This is a vision, ultimately, in which patterns and flows of carbon metabolism—between the atmosphere, the biosphere and the commodity realm—are deliberately reconfigured, in ways argued to be beneficial to the climate system, through a collaborative division of labour in which trees serve principally to remove carbon from the atmosphere, while humans serve principally to emplace that carbon in commodity forms and storage facilities located elsewhere.[92]

The emergence of forest-based bioenergy production in the US South is therefore significant not merely for the role that it plays, alongside the liquid transport biofuel sector discussed in Chapter 2, in the ongoing spatial restructuring of global energy–society relations. More significantly, wood pellet manufacturing reveals the productive potential for modern bioenergy to act as a means of creating entirely new forms of economic value from plants, specifically by enrolling the work of vegetal lifeforms—in this case, loblolly pine trees—not just in the practices of energy resource-making, but more ambitiously in efforts to deliberately alter the composition of the global atmosphere. Yet the case also reveals how bioenergy logics, in seeking to generate these new forms of value, are predicated on the strategic disaggregation of the diverse vital capacities that vegetal natures possess. In working forests across much of the US South, where the goal is not to store carbon dioxide but rather convey it to other places, this means that no loblolly pine is ever regarded as indispensable; rather, each individual tree exists merely to be replaced—ideally as quickly as possible—by another. Trees themselves are hereby transformed, in these forests, into forms of nature whose value to society depends fundamentally on their being consumed, whether as bioenergy or indeed through other means. The implications of this strategic disaggregation of the capacities of vegetal life to photosynthesise and grow from the capacities of vegetal life to bring energy to rest as matter will be discussed further in Chapter 5, which examines expanding efforts to combine bioenergy with carbon capture and storage technologies, effectively substituting naturally occurring forest carbon sinks with deliberately engineered, subterranean ones. Before moving on to this case, however, Chapter 4 first turns the book's attention to the question of modern bioenergy's implications for the production of new kinds of vegetal nature, focusing on efforts to develop bespoke, high-yielding and dedicated energy crops for the global aviation sector.

Notes

1 Drax Group plc, "Drax's largest biomass shipment arrives at the UK's biggest biomass handling facility."

2 Only twelve days after the arrival of *Zheng Zhi*, a subsequent shipment, again totalling more than 60,000 tonnes of pellets, became the one hundredth to arrive at Immingham from Baton Rouge in a period of just five years. See Drax Group plc, "Drax receives 100th biomass cargo from dedicated US export facility."

3 BBC News, "Britain goes coal free as renewables edge out fossil fuels."

4 See for example Ambrose and Kommenda, "Britain breaks record for coal-free power generation"; and Cockburn, "UK's record coal-free power run comes to an end."

5 US Department of Agriculture Foreign Agricultural Service, *Global Agricultural Information Network Report: UK Wood Pellets Annual 2024*.

6 Drax Group plc, *Committed to the World's Energy Transition*.

7 Between 2012 and 2024, Drax Group plc, owner of Drax Power Station, received an estimated total of £7 billion through policy mechanisms intended to incentivise biomass burning, including the UK Renewables Obligation and a bespoke "Contract for Difference" guaranteeing a minimum income for each megawatt hour (MWh) of electricity generated by burning wood pellets. See Pratley, "Drax is the subsidy show that goes on and on."

8 For a detailed review of policy mechanisms used to promote biomass burning internationally, see Roni *et al.*, "Biomass co-firing technology with policies, challenges, and opportunities."

9 In 2023, 73 per cent of all wood pellets imported to the UK were sourced from the US (see US Department of Agriculture Foreign Agricultural Service, *Global Agricultural Information Network Report: UK Wood Pellets Annual 2024*). As of 2017, nearly three quarters of US wood pellet manufacturing capacity was located in the South region (Oswalt *et al.*, *Forest Resources of the United States, 2017*, 47).

10 On "vegetal labour" see Palmer, "Putting forests to work?"

11 Demeritt, "Scientific forest conservation and the statistical picturing of nature's limits in the Progressive-era United States," 438.

12 Prudham, *Knock on Wood*, 96; see also Michael Williams, *Americans and Their Forests*.

13 Prudham, *Knock on Wood*, 95.

14 Hodges *et al.*, "Recession effects on the forests and forest products industries of the South."

15 Oswalt and Smith, *U.S. Forest Resource Facts and Historical Trends*.

16 Oswalt *et al.*, *Forest Resources of the United States, 2017*.

17 Oswalt *et al.*, *Forest Resources of the United States, 2017*, 93.

18 Oswalt *et al.*, *Forest Resources of the United States, 2017*, 9.

19 The intensity of the commercial activity shaping forests of the US South has been evidenced by satellite data showing rates of forest disturbance four times greater, between 2000 and 2012, than those prevailing in South American rainforests. Thirty-one per cent of all forests in the US South were either lost or regrown in this period (Hansen *et al.*, "High-resolution global maps of 21st-century forest cover change").

20 Oswalt *et al.*, *Forest Resources of the United States, 2017*, 4; and interview, North Carolina Forest Service official, May 2019.

21 Hitchner *et al.*, *Forests as Fuel*, 192.

22 Wear and Greis, *The Southern Forest Futures Project*; and Forest2Market, *Historical Perspective on the Relationship between Demand and Forest Productivity in the US South*.

23 For a fuller discussion of industrial tree improvement programmes, focusing especially on the forests of the Pacific coast region, see Prudham, "Taming trees."

24 McKeand *et al.*, "Performance of improved genotypes of loblolly pine across different soils, climates, and silvicultural inputs."

25 Oswalt *et al.*, *Forest Resources of the United States, 2017*, 28.

26 Boyd *et al.*, "Industrial dynamics and the problem of nature," 557.

27 Drax Group plc, "How wood pellets are revitalizing a community."

28 Southern Environmental Law Center, *Pellet Mill Community Impact Survey*, 3; and Voegele, "USDA: US wood pellet exports top 926,024 metric tons in May."

29 See for example Schlesinger, "Are wood pellets a green fuel?"

30 Southern Environmental Law Center, *Pellet Mill Community Impact Survey*, 1.
31 Koester and Davis, "Siting of wood pellet production facilities in environmental justice communities in the southeastern United States," 70.
32 Paragraph 13 of Annex VI of the EU's "recast" Renewable Energy Directive, passed into law in 2018, declares that "emissions of CO_2 from fuel in use ... shall be taken to be zero for biomass fuels" (see European Commission, "Directive (EU) 2018/2001," 185).
33 Sterman *et al.*, "Does replacing coal with wood lower CO_2 emissions?"
34 Chatham House, *Woody Biomass for Power and Heat*, 5.
35 Iuel-Stissing *et al.*, "Biomass qualification and the dynamics of market framing," 177.
36 Sterman *et al.*, "Does wood bioenergy help or harm the climate?" 134–135.
37 Searchinger *et al.*, "Europe's renewable energy directive poised to harm global forests."
38 Beddington *et al.*, "Letter from scientists to EU Parliament regarding forest biomass." See also Stephenson and MacKay, *Life Cycle Impacts of Biomass Electricity in 2020*; and Matthews *et al.*, *Carbon Impacts of Biomass Consumed in the EU*.
39 Dogwood Alliance *et al.*, *European Imports of Wood Pellets for "Green Energy" Devastating US Forests*; see also Hitchner *et al.*, *Forests as Fuel*.
40 Dogwood Alliance, "Why is biomass bad?"; see also Juskus, "Sacrifice zones."
41 Vickers, "Clearing the air on how wood pellet factories affect their neighbors and the climate."
42 See Environmental Integrity Project, *Dirty Deception*, 30.
43 Hitchner *et al.*, *Forests as Fuel*, 110.
44 Hitchner *et al.*, *Forests as Fuel*, 110.
45 See for example Southern Environmental Law Center, *Pellet Mill Community Impact Survey*.
46 Davies, "Slow violence and toxic geographies," 421; see also Nixon, *Slow Violence and the Environmentalism of the Poor*.
47 Davies, "Slow violence and toxic geographies," 409.
48 Cunningham *et al.*, *Managing Loblolly Pine Stands ... from A to Z*.
49 In the words of one representative of a bioenergy corporation actively operating in the forests of the US South, a considerable amount of work is required simply "to hold things in this pine position".
50 Prudham, *Knock on Wood*, 102.
51 Interview, wood pellet industry representative, May 2019.
52 Interview, forest product industry representative, May 2019.
53 White, "Forest owners' boss: 'Sustainable forest management crucial to maintaining carbon cycle'."
54 Simon, "Brussels rebuked over 'confusing' draft EU bioenergy rules."
55 Wear and Greis, *The Southern Forest Futures Project*, 2.
56 Forest2Market, *Historical Perspective on the Relationship between Demand and Forest Productivity in the US South*, 6.

57 Drax Group plc, "Active management of forests increases growth and carbon storage."
58 Boyd *et al.*, "Industrial dynamics and the problem of nature," 566.
59 Since they reduce stand density and provide opportunities to eradicate unhealthy trees, thinning operations are also sometimes credited with reducing the forest's susceptibility to wildfire and to insect-borne disease. See Cunningham *et al.*, *Managing Loblolly Pine Stands ... from A to Z.*
60 Cunningham *et al.*, *Managing Loblolly Pine Stands ... from A to Z.*
61 Forestry Commission England, *Sustainable Forest Management*, 9.
62 Baker and Langdon, "*Pinus taeda.*"
63 Interview, wood pellet industry representative, April 2018.
64 Interview, wood pellet industry representative, April 2018.
65 Peltola and Tuomisaari, "Making a difference: Forest biodiversity, affective capacities, and the micro-politics of expert fieldwork."
66 Interview, wood pellet industry representative, April 2018.
67 Interview, forest product industry representative, September 2018.
68 Interview, forestry professional, May 2019.
69 Interview, wood pellet industry representative, April 2018.
70 Boyd *et al.*, "Industrial dynamics and the problem of nature," 564.
71 See Demeritt, "Scientific forest conservation and the statistical picturing of nature's limits in the Progressive-era United States."
72 Boyd *et al.*, "Industrial dynamics and the problem of nature," 566.
73 Schelhas *et al.*, "What family forest owners talk about when they talk about trees," 6.
74 It is critical to reiterate that these are relative measures. Some studies contend that larger, older trees sequester greater absolute volumes of carbon dioxide than younger, ostensibly more vigorous trees each year, simply because of their size. See for example Moomaw *et al.*, "Intact forests in the United States."
75 Drax Group plc, "What is a working forest?"
76 On the idea that nature-based production systems can serve to establish such a moral economy, see Paxson, "The naturalization of nature as working."
77 For a more sustained discussion, see Palmer, "Putting forests to work?"
78 Interview, bioenergy industry representative, September 2018.
79 This effect is particularly accentuated where those wood products end up being used in place of steel, concrete or other materials which themselves require significant amounts of energy to produce. See Oliver *et al.*, "Carbon, fossil fuel, and biodiversity mitigation with wood and forests."
80 Moomaw *et al.*, "Intact forests in the United States."
81 Lewis *et al.*, "Regenerate natural forests to store carbon."
82 See Simard, *Finding the Mother Tree.*
83 Hitch and Grignon, "A forest of energy," 252.
84 Palmer, "Putting forests to work?"
85 In the UK context, the attachment of these additional forms of value to wood pellets is indicated by the significant government financial incentives provided to energy companies converting from coal to biomass burning in the past decade (see note 7 to this chapter).

86 Drax Group plc, "5 things you never knew about forests."
87 Interview, forest product industry representative, May 2019.
88 Davies, "Slow violence and toxic geographies"; and Southern Environmental Law Center, *Pellet Mill Community Impact Survey*, 1.
89 See Daggett, *The Birth of Energy*.
90 Paxson, "The naturalization of nature as working."
91 Besky and Blanchette, *How Nature Works*, 4.
92 This idea of a collaborative division of labor echoes Alyssa Battistoni's concept of "hybrid labor" (Battistoni, "Bringing in the work of nature"). See also Ernwein, "Bringing urban parks to life."

4

Living oil fields: reprogramming crops to mine the sun

Decarbonising aviation without the use of biofuels is hardly imaginable.

—Mission Possible Partnership, *Making Net-Zero Aviation Possible*, 2022

Mark my words—the next 20 years, farmers are going to be providing 95 percent of all the sustainable airline fuel.

—Former US President Joe Biden, 2023

On 28 November 2023, a Boeing 787 touched down at New York's JFK Airport having completed the first transatlantic flight by a large passenger aircraft to be powered exclusively by so-called "sustainable aviation fuel" (SAF)—an alternative to conventional jet fuel derived not from fossil oil, but from biological feedstocks. On commencing its 3,500-mile journey at London's Heathrow Airport several hours earlier, the aircraft had been laden with approximately 50 tonnes—or around 60,000 litres—of SAF, derived in this case from a combination of "waste fats ... and wastes of corn production from the US".[1] While the flight carried no fare-paying passengers, it was nonetheless hailed by the commercial airline which operated it as a "significant milestone on the path to decarbonising aviation".[2] One of the dignitaries on board, then-UK Secretary of State for Transport Mark Harper, remarked loftily upon landing that "history has been made".[3] Whether historic or not, the flight certainly served an important symbolic function in light of high-profile commitments by the global aviation sector to radically reduce its climate change impacts over the coming decades. At its 2021 Annual General Meeting, the International Air Transport Association committed to achieving "net zero carbon emissions" in the global air transport industry by 2050, forecasting that 65 per cent of the sector's annual emissions—or around 1.2 gigatons of carbon—could be abated through the use of SAF by mid-century.[4]

Dreams of fully decarbonising air travel, however, will not be easy to get off the ground. Not only does the industry's embrace of SAF rely on contestable definitions of biomass-based fuels as inherently carbon neutral, despite the fact that the combustion of such fuels unequivocally does release GHGs into the atmosphere, as discussed in previous chapters. More precariously, these dreams rest on the assumption that a truly colossal volume of sustainable biomass feedstocks will be cultivated and funnelled into the production of renewable aviation fuel over the coming decades. Over the course of a typical transatlantic flight, a Boeing 787 burns approximately three litres of fuel per *second*. To replace even 10 per cent of the fuel forecast to be burned by the global aviation sector in 2030, as a coalition of sixty key industry players committed to do in September 2021,[5] would require approximately 40 million tonnes of SAF, or around 50 billion litres.[6] But this is to say nothing of the scale of fuel supply which would be required to help render global aviation hypothetically net zero in carbon dioxide terms by 2050.[7] According to a 2022 report into the prospects of "making net-zero aviation possible", led by the World Economic Forum (WEF) and American consultancy giant McKinsey and Company, and backed by over sixty companies from across the aviation sector, around 220 million tonnes of bio-jet fuel—equivalent to approximately 275 billion litres—would be needed to supply just 50 per cent of the sector's final energy demand in 2050.[8] Without drastic growth in the production of other kinds of alternative fuels, most notably from hydrogen and electricity, the sector's demand for sustainable biomass feedstocks could end up being even higher still. To say then that the projected growth trajectory for bio-jet fuel over the coming years is ambitious would, in short, be putting things extremely politely.

Yet in 2023, despite the fact that worldwide SAF production doubled against figures from the previous year, total output still only reached 600 million litres, accounting for just 0.2 per cent of global jet fuel use.[9] While SAF can theoretically be derived from any biological material, including wastes like used cooking oil, residues and byproducts from existing agricultural or forestry operations, or even algae, the supply of sufficient volumes of feedstock to help achieve a net zero goal for the entire aviation sector will require nothing short of a global agricultural transformation. According to the WEF and McKinsey scenario outlined above, for example, producing 275 billion litres of bio-jet fuel every year would require anything between 10 per cent and one quarter of all the world's available sustainable biomass resource to be marshalled simply to help keep aircraft airborne.[10] So steep is the required increase in SAF supply, in fact, that efforts to reach what the UK government has glibly termed "jet zero" can be viewed as perhaps the single most influential policy driver undergirding a growing push to revolutionise the field of bioenergy production itself. After all, with the future of flying apparently now at stake, relying on the conventional processing of staple food crops into bioethanol or biodiesel, as has typically occurred in the production of biofuels for road transport (see Chapter 2), is unlikely to cut the mustard. Instead, policymakers are directing significant sums of public money into research which aims to develop more advanced forms of bioenergy—wherein genetic engineering is used to actively optimise both the raw productivity and the underlying physiological characteristics of dedicated energy crops, enabling both larger absolute volumes and larger proportions of plant biomass to be converted into combustible fuel. Indeed, as the chapter will argue, an apparently shared goal of a growing array of scientific institutes and commercial enterprises active in the realm of advanced bioenergy is to engineer key crop species to produce hydrocarbon molecules which, even in the field, are already compatible with prevailing fuel combustion technologies. The aim, in short, is to render plant growth itself increasingly indistinguishable from oil provisioning.

Focusing on the case of the US specifically, this chapter will show how imaginaries of a net zero future for the aviation sector—and indeed of a future bioeconomy writ large—hinge on the enrolment of abundant, fast-growing, genetically engineered crops and trees to depoliticise productivist logics of energy generation and use, ultimately by producing new kinds of vegetal nature which enact productivism at the molecular level. The chapter begins by introducing a key regulatory driver of these transformations—a so-called Sustainable Aviation Fuel "Grand Challenge Roadmap" launched jointly by several US federal government agencies in 2022—placing this endeavour into the context of wider efforts to kickstart a wholesale shift to a sustainable American bioeconomy. It then moves on to examine the forms of expert knowledge, representation and calculation being used to help bring a putative sustainable crop bioenergy resource into being, understood as a regulatory object ostensibly amenable to monitoring, control and indeed optimisation and improvement at the national scale. Having reflected on the implications of these resource-making processes for understandings both of agricultural land and of wider relationships between humans and plants in rural America, the chapter then moves on to scrutinise scientific engagements with plant life in research fields dedicated to the production of advanced bioenergy feedstocks, including for SAF. More specifically, the chapter examines widespread scientific conceptualisations of plants as sources of potentially limitless productivity and versatility, wherein plant bodies come to be viewed as containers within which the generation of diverse bio-commodities can be staged on demand. Against this view of plants as sites of work, and quite literally in some cases of "plants as factories",[11] the chapter concludes by calling for a more explicit recognition of the plants enrolled in this drive for advanced bioenergy and large-scale SAF production as sources of labour power generating surplus value in their own right—in short, for a view of plants not as sites of work, but as workers themselves.

Cleared for take-off: sustainable aviation fuels and the American bioeconomy

In September 2022, the US Departments of Energy, Transportation and Agriculture, working in tandem with the Environmental Protection Agency, jointly launched the SAF Grand Challenge Roadmap aiming to "reduce the cost, enhance the sustainability, and expand the production and use of Sustainable Aviation Fuel ... to meet 100% of demand by 2050".[12] This Grand Challenge applies to all domestic aviation fuel "uplift" in the US—a measure of the total amount of jet fuel loaded onto aircraft at US airports each year, including domestic passenger and cargo flights, as well as international departures operated by American and foreign airlines.[13] In 2019, jet fuel uplift in the US totalled approximately 23 billion gallons (or 87 billion litres), but the figure is expected to grow in the coming decades, eventually reaching approximately 34 billion gallons (or just under 130 billion litres) by the middle of this century.[14] Producing such a vast amount of SAF domestically by 2050 would be no mean feat, but arguably just as ambitious—given that a mere five million gallons of SAF were produced in the US in 2021—is the shorter-term goal of producing three *billion* gallons per year (or around 11 billion litres) as soon as 2030. Even though US SAF production has been on an upward trajectory, reaching approximately 23 million gallons in 2023, this remains—as was noted by a recent National Renewable Energy Laboratory report—"more than two orders of magnitude" below the rapidly approaching 2030 target.[15]

To be clear, no US government agency expects all of America's SAF requirements to be met through the conversion and combustion of dedicated energy crops, or what might also be termed "purpose-grown biomass resources".[16] Significant investments will also be made in research intended to help achieve the production of large volumes of SAF from agricultural and forestry wastes, algae, otherwise unmarketable woody biomass derived from forests, and even potentially carbon dioxide captured from emissions generated by other industries. If SAF production is to be drastically ramped up by a factor of more than one hundred in the short period between 2023 and 2030, however, government analysis suggests that 90 per cent of this increase will come from so-called "sustainable lipid" feedstocks,[17] of which the most readily and widely available—at least in the US—are purpose-grown oilseed crops such as soybean and canola. Speaking in July 2023, then-US President Joe Biden was adamant about the critical role to be played by American agriculture in facilitating the transition to net zero aviation, arguing that in the "next 20 years, farmers are going to be providing 95 percent of all the sustainable airline fuel".[18] There are strong echoes, then, in this new American rush for SAF, of the enthusiasm which was shown by political leaders for transport biofuels in the early 2000s—as a means both of supporting American farmers and of boosting energy security and independence (as set out in Chapter 2). Indeed, while the policy focus may have shifted skywards, away from driving and on to flying, "silver bullet"-style framings of biofuels as a solution to environmental, economic and political goals remain pervasive and potent in today's America. To quote from a 2021 Memorandum of Understanding signed by representatives of the US Departments of Energy, Transportation and Agriculture on SAF, for example, efforts to increase the production of these fuels should be viewed as "an investment in America that not only reduces our environmental impact, but also supports energy independence and creates jobs in agriculture, forestry, infrastructure, research and development and other areas where America already excels at production".[19]

Beyond the strategic entanglement of environmental goals with agendas relating to rural jobs, economic growth and energy independence, similarities also persist in the extent to which biofuels for the aviation sector are depicted not only as straightforward substitutes for the petroleum-based fuels they are intended to replace, but also as complementary to—rather than conflicting with—fundamental agricultural priorities of food and animal feed production. Thus, not only does the Grand Challenge Roadmap insist that SAF "is 'drop-in' liquid hydrocarbon jet fuel ... compatible with *existing* aircraft and engines",[20] but it also talks up the potential for large volumes of these fuels to be derived from "dedicated feedstock production on marginal and degraded lands".[21] If anything, hopes around this latter set of potentials have only grown since the early 2000s, with large-scale SAF production today invoked not only as a basis for enabling farmers to generate additional income from previously unproductive or unviable land, but also as a means of achieving environmental remediation, for instance by incentivising crop growth on land damaged by strip mining, or even by providing a marketable use for excess wood presently rendering many American forests vulnerable to wildfire and disease.[22] Taken in the round, these perspectives reinforce a view of bioenergy's "downstream applications" as seamless with existing, largely fossil fuel-based energy systems, while at the same time downplaying what are likely to be "significant upstream disruptions" associated with land-use changes, changing land ownership patterns and of course the establishment of considerable new industrial facilities for the processing of biomass fuels themselves.[23] But these ideas are far from newfangled; the appeal of so-called "drop-in" biofuels, for example, has been widely noted since the early 2010s.[24] Similarly, views of bio-jet fuel production as a complement—and not a threat—to food production can be traced back to some of the earliest SAF trials in North America; in 2012, for instance, following a successful ninety-minute business jet flight in Canada powered solely by fuel derived from the oilseed crop *Brassica carinata*, industry representatives boasted that "by growing this seed you actually enhance or improve the land for eventual food production".[25]

Despite this apparent sense of déjà-vu, however, there have been important shifts in American discourses around large-scale modern bioenergy since the early 2000s' new biofuel rush. For one thing, many concerns about the sustainability and ecological impacts of intensive crop cultivation for the bioenergy market—previously the preserve of environmental campaign groups and other critical voices outside of government—have now gained credibility in official policymaking circles as well. Indeed, American efforts to meet 100 per cent of aviation fuel demand with SAF in 2050 are conditional on the development of fuels which can claim to reduce GHG emissions compared to fossil-based jet fuel by at least 50 per cent, importantly including an assessment of the estimated ILUC effects induced by cultivating feedstocks on the ground.[26] Concerns over potentially excessive groundwater use are also deemed to be a "critical component" of the sustainability question—not least given the relatively arid climatic conditions experienced by many of America's most productive agricultural states.[27] Both issues—ILUC and groundwater depletion—have turned out to be bad news for farmers and industry protagonists hoping to see generous government support for SAF production from crops like corn and soy; as recently as April 2024, the Department of the Treasury insisted that corn-based SAF could only qualify for tax credits if a "bundle" of so-called "climate-smart agriculture" practices were also adopted by farmers, including the use of no-till soil management, cover crops to reduce soil erosion and a switch to more efficient fertilisers.[28] In short, government attitudes towards the question of how to do bioenergy production well have become more discerning, leaving staple crops like corn and soy with an uphill battle to prove themselves as worthy feedstocks for SAF production. Certainly, former President George W. Bush's glib idea that any future US president might regularly receive a "crop report" outlining the contributions of those two species—corn and soy—to reducing national levels of fossil fuel use, and hence to improving American energy independence, seems now to be dead in the water.[29]

More fundamentally still, and again in a manner that marks a step change from ideas surrounding the crude conversion of staple food crops into road transport biofuel in the early 2000s, contemporary visions of a gargantuan SAF industry are much more intricately tethered into wider US policy agendas pushing, at least rhetorically, for a wholesale structural shift in the American economy—away from fossil resources and towards biological ones. This is the idea, simply put, of harnessing the power of biotechnology to create a "sustainable, safe, and secure American bioeconomy".[30] In a press statement issued in late 2022, the US White House depicted biology itself as a source of almost limitless possibilities for novel commodity production and value generation, all capable of being unlocked through the application of cutting-edge biotechnology and biomanufacturing techniques. We will "be able to realize", the statement contends, "the potential of biology to make *almost anything* that we use in our day-to-day lives, from medicines to fuels to plastics, and continue to drive US innovation into economic and societal success".[31] Among the key components of this vision of a practically limitless bioeconomy is so-called "advanced" bioenergy—a burgeoning field of research and development aiming not simply to redirect existing agricultural outputs to new ends, but also to use biotechnology to actively optimise and reconfigure a wide array of dedicated energy crops to help meet diverse environmental, societal and economic needs. Indeed, targets for the domestic production of SAF feature prominently in a 2023 White House report setting out a series of "bold goals" for the US bioeconomy more broadly.[32]

Accordingly, even while the SAF Grand Challenge Roadmap concedes that considerable weight will need to be placed on readily available oilseed crops to reduce emissions from the aviation sector in the next decade or so, any hopes of producing well over 100 billion litres of SAF annually by the middle of this century will inevitably require the development, through genetic engineering, of more specialised, efficient and productive energy crops altogether. As will be discussed later in this chapter, much of this research agenda focuses not on oilseed crops at all, but rather on high-yielding herbaceous grass species such as switchgrass, *Miscanthus* or even sugarcane, where genetic enhancements are already being trialled to help achieve "engineered oil excrescence in biomass itself".[33] In this last sense at least, the goal of plant science dedicated to realising a "sustainable, safe, and secure American bioeconomy" can literally be to turn plants historically cultivated for their starch and sugar content into crops cultivated for a different commodity altogether—oil itself.[34] Before examining scientific engagements with plant life in research fields dedicated to these objectives, however, the chapter first moves on to scrutinise the forms of expert knowledge, representation and calculation being used to help bring a putative sustainable crop bioenergy resource into being in America today.

Let a billion tonnes of biomass boom! Making America's advanced bioenergy resource

Despite the sheer scale of growth in domestic SAF production being targeted by US federal government agencies today, 2022's Grand Challenge Roadmap is bullish about America's ability to supply sufficient biomass resources to help realise this vision. "About 1 billion dry tons/year of biomass can be grown or collected sustainably", the roadmap contends, emphasising that such a vast amount of biomass could easily be converted, according to official US Department of Energy (DOE) estimates, into "50–60 billion gallons of advanced biofuels without impacting agriculture, trade, or current uses of biomass".[35] The source of this rosy outlook, intriguingly, is a report whose very title makes the goal of achieving a one-billion-ton sustainable biomass haul from domestic US resources entirely explicit.[36] In fact, the *2016 Billion-Ton Report: Advancing Domestic Resources for a Thriving Bioeconomy*—led by analysts at Tennessee's Oak Ridge National Laboratory—marks the third instalment in a series of so-called "Billion-Ton Reports" sponsored by the DOE, dating back to 2005, when then-President George W. Bush was waxing lyrical about a bright future for "home-grown" biofuels as alternatives to conventional gasoline on America's roads.[37] In early 2024, the series was extended again with the publication of the *2023 Billion-Ton Report: An Assessment of U.S. Renewable Carbon Resources.*[38] While the title of this latest iteration has shifted subtly, omitting reference to a "thriving bioeconomy" in favour of a framing of biomass itself as a form of "renewable carbon", it nonetheless continues to pre-empt the report's overall conclusions, which hence contain few surprises. Indeed, depending on the strength of overall demand, the 2023 report contends that America will be capable of producing anywhere between 1 and 1.5 billion tons of biomass in a "mature-market" scenario, all without jeopardising the country's ability to meet "projected demand for food, feed, fiber, and exports".[39] Even on the lower end of these estimates, the latest report underscores that future sustainable domestic biomass supply on this scale would be sufficient to produce "1.7 times the quantity [of fuel] needed to achieve the SAF Grand Challenge".[40] The report's authors insist, however, that their analysis is "policy agnostic, end-use agnostic, not prescriptive, and not predictive".[41]

While the DOE's Billion-Ton Reports may talk up the potential for a drastic expansion in the size of America's sustainable bioenergy resource, this will not bring about increased SAF production by itself. The *2016 Billion-Ton Report*, for example, drew explicit attention to a now-defunct US Federal Aviation Administration goal of reaching one billion gallons (or just under four billion litres) of alternative jet fuel production by 2018, a target whose scale appears almost absurdly over-ambitious when set against the mere five million gallons (or 19 million litres) of SAF which was actually produced a few years later in 2021.[42] Even if these reports neither can nor do claim to be predictive, however, simply envisaging a prodigious growth in sustainable biomass supply, one sufficient to "approximately triple" the size of the US bioenergy economy as of 2022,[43] already offers an important discursive foundation for protagonists of large-scale modern bioenergy as a viable alternative to fossil fuel use across a range of settings, including organisations promoting SAF as key to achieving net zero carbon emissions in aviation. From the vantage point of scholarship in science and technology studies, these assessments can be viewed as a classic "technique of futuring"—a means not simply of providing policymakers with information arising from "neutral investigation", but rather of performing a specific, detailed image of the future, thereby enabling diverse political actors to converge around shared visions of what forms of action and intervention are feasible and appropriate.[44] Writing in the context of expert assessments of future climate change scenarios, Silke Beck and Martin Mahony contend that such techniques do not simply offer up "useful information"; instead, they effectively perform a "world-making" function by "furnishing policymakers with the objects, variables and relations upon which they seek to exert influence".[45] From this standpoint, then, the approach that the DOE's Billion-Ton Reports take to envisaging a radically expanded, sustainable domestic biomass supply for the US—including the accounts that these reports give of the objects, variables and relationships which must allegedly be optimised if such an expansion is to be realised—warrants much closer scrutiny.

At an overarching level, any analysis of future biomass production potential must grapple with a vast array of pathways for obtaining biomass itself. The DOE's *2023 Billion-Ton Report*, accordingly, offers in-depth assessments of the potentials for increased volumes of biomass to be sourced not only from forested lands and cropland, but also from various waste streams and a number of so-called "emerging" resources (including microalgae, macroalgae and even point-source emissions of carbon dioxide). Beyond simply offering a supply-side analysis of the biophysical and technical constraints likely to impinge upon increased biomass production rates within each of these various pathways, the DOE's assessments also draw heavily on socioeconomic models to illustrate how supply itself might be shaped by evolving future demand for a range of different biomass feedstock types. The most recent report, for example, not only aims to model the scale of America's future sustainable biomass resource under "near-term" and "mature-market" scenarios, but also distinguishes between mature-market scenarios in which demand for biomass is presumed to offer a low, moderate or high degree of "pull" on supply.[46] Strikingly, however, while this 2023 report shows appreciable increases in biomass production from America's waste, forest and agricultural sectors between "near-term" and "mature-market" scenarios, readers seeking to compare "mature-market" scenarios characterised by low, medium and high levels of demand will find increases in supply located exclusively in the agricultural realm (see Table 4.1). Indeed, while the DOE's analysis suggests that—regardless of demand—forestlands

Table 4.1 Projected US domestic supply of sustainable biomass feedstocks from waste, forestry and major agricultural streams (in millions of dry tons per year).

Feedstock source	Current supply	Near-term increase	Mature-market increase (low)	Mature-market increase (medium)	Mature-market increase (high)
Wastes	37	179	217	217	217
Forestlands	143	30	63	63	63
Energy crops (herbaceous)	162	0	284	345	535
Energy crops (woody)		0	34	54	103
Agricultural residues		138	153	179	201
Total supply	**342**	**689**	**1,093**	**1,200**	**1,461**

Note: Figures shown correspond to a scenario which models biomass prices at $70 per dry ton. Government information at US Department of Energy (DOE) websites is in the public domain (see https://www.energy.gov/web-policies).

Source: Author's table, derived from data presented in the US DOE's *2023 Billion-Ton Report.*

will be able to provide no more than 63 million extra dry tons of biomass in a mature-market scenario,[47] and similarly waste streams will offer up no more than an additional 217 million dry tons,[48] outputs from agricultural land are by contrast deemed to have the capacity to expand prodigiously. Even in a low-demand scenario, a mature market is already estimated to have the potential to induce American farmers to supply around 471 million dry tons of additional biomass when compared to current levels. Two thirds of this additional total would be derived, the report suggests, from dedicated energy crop production, with the remainder being sourced from the increased recovery of so-called crop residues—in essence, the stalks and leaves of conventional food crops, such as corn and wheat, which are normally left over after harvesting. Under conditions characterised by medium and high levels of demand, however, the 2023 report depicts a future US agricultural sector capable of furnishing bioenergy markets with even higher additional supplies of feedstock—to the tune of 578 and 839 million dry tons of biomass, respectively.[49]

Despite its stated "policy agnostic" and "not prescriptive" outlook, then, the DOE's latest Billion-Ton Report nonetheless conveys a clear message to readers about the potential for targeted interventions—aimed at agriculture specifically—to maximise the scale of US sustainable biomass supply over the coming decades. The cultivation of dedicated energy crops, in particular, appears to offer exceptional scope for growth; in a mature-market scenario exhibiting medium levels of demand, the report suggests that additional output from herbaceous energy crops alone—relying chiefly on species like *Miscanthus*, switchgrass and sorghum—could more than double America's 2022 total biomass supply of just over 340 million dry tons.[50] The contention that dedicated energy crop cultivation on agricultural land could be expanded to such a radical extent *without* compromising America's ability to meet future demands for food, animal feed or indeed crop exports, however, relies on a number of critical assumptions in the socioeconomic models used by the *2023 Billion-Ton Report* to project future supply. Among the most notable is a presumption of annual increases in the dedicated energy crop yields through to at least 2040 of 1 per cent per year in a medium-demand scenario, rising to 3 per cent per year under conditions of high demand. While the report's model does not restrict energy crop cultivation only to so-called marginal lands deemed unsuitable for conventional crops at present, its outputs also suggest that regions of "prime farmland", such as in the so-called "corn belt" of the Midwest, would largely not be converted to purpose-grown energy crops "because of economic comparative advantages of commodity crops in that region".[51] Instead, dedicated energy crop cultivation would, according to the model, predominantly occur in "regions of less intensive conventional crop production", with production arising either from the conversion of livestock pasturelands or from less profitable tranches of existing cropland.[52] Where pastureland is converted to grow bioenergy crops, intriguingly, the report's model also presumes that landowners will not reduce livestock production; instead, management of remaining pasturelands is intensified through "timed rotations using paddocks that increase sustainable stocking density".[53] Overall, by subscribing to these modelling assumptions, the report paints a picture of dedicated energy crop cultivation as a "win-win" for America's energy and agriculture sectors, arguing that "the introduction of perennial purpose-grown energy crops can be a net energy resource for the nation while also playing a role in crop diversification, soil conservation, and improving farm income".[54]

Thus far, the focus has been on how the DOE's reports model the scale of future sustainable biomass output that could potentially be achieved by America's agricultural sector, rather than the spatial footprint of biomass cultivation itself. At an overarching level, and by subscribing to the assumptions regarding crop yield and land conversion preferences outlined above, the *2023 Billion-Ton Report* suggests that nearly 400 million dry tons of biomass could be supplied from using just 9 per cent of US agricultural land, or around 300,000 square kilometres, to grow dedicated energy crops.[55] Beyond offering aggregated estimates of the total amount of US land converted to purpose-grown energy crop cultivation under various market scenarios, however, the DOE's reports also seek to project the geography of future biomass production on a more fine-grained level. Indeed, from a spatial standpoint, the reports go far beyond offering assessments of the broad agricultural regions from which sustainable biomass supplies might be derived in the future. The latest report's underlying socioeconomic model, for example, despite relying on national-level estimates of future demand for bioenergy, food, animal feed and crop exports, estimates and models actual land-use allocations at the sub-state (county) level, aiming—within each of America's more than three thousand counties—to achieve the "most profitable allocation of land from the landowner's perspective".[56] By mapping these county-level projections, the DOE's Billion-Ton Reports offer readers compelling and ostensibly high-resolution maps of the potential geography of America's future sustainable biomass resource. A complementary online data portal, hosted within the DOE's publicly accessible "Bioenergy Knowledge Discovery Framework", even enables users to generate their own maps of future biomass outputs from distinct agricultural pathways.[57] One such map is shown in Figure 4.1, a projection of the spatial

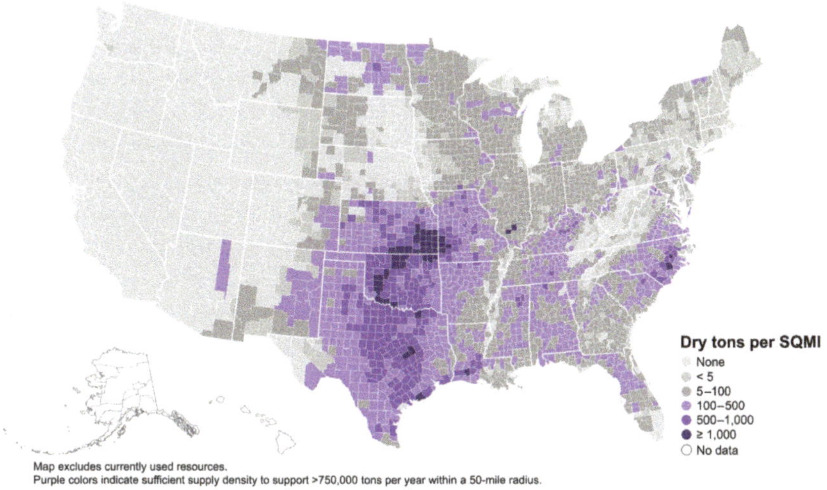

Figure 4.1 Projected rates of US dedicated herbaceous energy crop production at the county level under a "medium-demand, mature-market" scenario, measured in dry tons per square mile.
Source: Derived from the "Billion-Ton 2023 Data Portal" of the US Department of Energy's "Bioenergy Knowledge Discovery Framework", https://bioenergykdf.ornl.gov/bt23-data-portal.

distribution of the 345 million dry tons of biomass forecast to arise, under a mature-market scenario characterised by medium levels of demand, from dedicated herbaceous energy crop cultivation.

From the standpoint of scholarship in critical resource geography, the DOE's Billion-Ton Reports can be viewed, I argue, as classic tools of state-led resource-making, drawing as they do on "technical devices of measurement and visualisation" to effectively constitute America's future advanced bioenergy resource as a coherent entity amenable to future governance and optimisation.[58] With regard to biomass supplies envisaged to emerge from agricultural land specifically, the reports privilege particular calculative and metrological regimes—those of agricultural economics and land-use modelling—depicting landowners as rational economic actors seeking to maximise farm profits by weighing up trade-offs between land suitability, potential yield increases per year, and commodity prices as commanded "at the farm gate". By foregrounding and rendering legible this particular set of "objects, variables, and relations",[59] the reports enact what James Scott would have termed a "narrowing of vision", bringing "into focus certain limited aspects of an otherwise far more complex and unwieldy reality".[60] One notable omission from this outlook is the potential future impact on American farming of environmental change, including the process of global warming which large-scale modern bioenergy production is itself meant to remedy.[61] Also obscured, moreover, are the rich, place-specific histories of farming understood not merely as an economic endeavour but also as a social and cultural practice, one shaped among other things by family dynamics, community relationships, land ownership regimes and vernacular understandings of responsible and productive soil management or simply farm stewardship more broadly. Moreover, as part of this narrowing of vision, diverse crops—such as corn, sorghum, sugarcane, *Miscanthus* and switchgrass—are recognised merely as "feedstocks" offering up various bio-physical, technical and economic potentials to an expanded US bioenergy sector, rather than as plant species with their own unique, place-specific environmental histories and indeed cultural associations.[62] In short, this standpoint envisages the end products of bioenergy resource-making—ostensibly sustainable, "drop-in" hydrocarbon fuels—as entirely decoupled and alienated from their diverse, complex origins in place-specific, socio-vegetal-ecological relationships.[63]

The DOE's Billion-Ton Reports, however, are characterised not simply by their reliance on the calculative and metrological regimes of agricultural economics and land-use modelling, but also by their deployment of specific representational techniques of cartography and mapping. Visual representations such as that shown in Figure 4.1—a projection of modelled, county-level outputs of biomass onto an unlabelled territorial map of the US—can be viewed as a form of what geographer David Demeritt, writing in the context of US forestry governance in the late nineteenth and early twentieth centuries, has called "statistical picturing".[64] By offering the appearance of "precision, fidelity, and realism", Demeritt argues that statistical picturing serves as a particularly effective means of disguising the fraught relationships and tensions between "representations and their objects".[65] The DOE's Billion-Ton Reports' biomass maps are, of course, qualitatively distinct from the first attempts to map America's national forest resource, such as William Brewer's famous map of 1874. For one thing, these maps do not correspond with an actually existing resource, but rather aim to depict the potential scope—as envisaged by a socioeconomic model—of a future resource, one as yet unrealised. Nonetheless, and despite being separated by 150 years, the two maps share important familial resemblances, being characterised not only by "visual simplicity ... combined with fine detail and mathematical precision", but also by the use of various shades of the same colour to indicate resource density.[66] The singular focus on projecting a spatially mediated measure of productivity, meanwhile—"dry tons per square mile"—acts to transform America's heterogeneous agricultural landscapes, again in Demeritt's terms, "into an apparently unified and calculable body", one potentially amenable to precise control and optimisation.[67] While the Billion-Ton Reports themselves make clear that assessments of future American biomass supply must inevitably consider complex interactions between evolving demand for bioenergy and fluctuations in wider markets, both national and international, for food, animal feed and livestock, the accompanying maps—by stark contrast—eradicate this context altogether. Instead, they offer a view of America's rural landscapes as a blank surface, or what Jesse Goldstein has called a "passive repository", one awaiting improvement through the more efficient and productive cultivation of biomass, itself understood as little more than fuel or energy in waiting.[68]

Every plant a factory? Technoscientific imaginaries of advanced bioenergy

Having examined how specific forms of expert knowledge, calculation and representation are enrolled within official US government reports to help envisage—and render governable—future American supplies of sustainable biomass, this section now moves on to address scientific engagements with plant life in research fields dedicated to the production of agricultural feedstocks for use as advanced biofuels, whether in the aviation sector or indeed beyond.[69] Under the auspices of its Genomic Science Program, the

DOE has since 2007 funded a vast amount of biological and environmental research dedicated to "overcoming the critical basic science challenges to cost-effective production of biofuels and bioproducts from plant biomass".[70] Much of this strategic research is coordinated and managed by four specialist Bioenergy Research Centers (BRCs), currently funded by the DOE to the tune of over $100 million per year.[71] In contrast to the Billion-Ton Reports and their high-level engagement with questions of supply and demand, commodity price signals and land-use trade-offs, scientists working in these DOE-sponsored BRCs focus principally on understanding how the intricate biological make-up of various crop plants might be manipulated and optimised for energy feedstock production. In practice, a great deal of this research aims either to increase total crop yields (often simply by maximising the production of plant biomass itself), to enhance the environmental resilience of key species—most notably to extreme heat and to drought—or finally to render plant biomass itself more readily amenable to conversion into desired biofuels and bioproducts. While these research challenges are framed by the DOE as "basic science", the realisation of more efficient and productive biofuels—what then-US Secretary of Energy Jennifer Granholm, speaking in 2021, called "one of our most promising paths to zero-carbon aviation"—must nonetheless confront several devilishly complex issues.[72]

Perhaps the most prosaic impediment to the straightforward derivation of larger volumes of hydrocarbon fuel from staple agricultural crops is the fact that the majority of the biomass contained in these crops takes the form neither of oil nor of sugar, but rather of lignocellulose—a fibrous, inedible material which makes up the majority of plant stalks and leaves. Being composed of carbohydrates and lignin, lignocellulose theoretically offers the bioenergy sector a major additional source from which to fashion combustible hydrocarbon compounds. Separating carbohydrates from lignin, however—perhaps unsurprisingly given the function that lignocellulose plays in providing robustness to plants at the level of individual cell walls—is far from easy. Indeed, plant science has long referred to lignocellulose as an inherently "recalcitrant" substance, unwittingly evoking the terminology used by critical resource geographers in more recent decades to describe forms of nonhuman nature not easily given to human control or optimisation.[73] Lignocellulosic recalcitrance has been a tenacious problem for those seeking to derive biofuels from crops in the US. Were the job of converting this portion of crop biomass into combustible fuel more straightforward, the almost forgotten "chemurgy" movement of 1930s and 1940s America might not have encountered such severe economic difficulties in its quest to usher in "the dawn of the alcohol era in motor fuels", for example.[74] In more recent times, a spate of investment in cellulosic biofuel research and processing plants in the US in the late 2000s and early 2010s has—in common with efforts in the EU—largely petered out, with many industrial facilities now in "idle or on-hold states" owing to "technical difficulties as well as high production costs".[75] In more recent years, scientists have increasingly turned away from the challenge of overcoming lignocellulosic recalcitrance, instead embracing the tools of biotechnology to find entirely new pathways for rendering agricultural crops into highly productive bioenergy resources.

One such pathway begins from the fundamental process through which plants convert solar energy into their own matter: photosynthesis. Indeed, many plant scientists working on the enhancement of agricultural crop productivity—whether funded by government agencies keen to help realise net zero aviation, or by philanthropists seeking to eradicate global hunger—share an underlying disdain for this process, at least as it has evolved in most of the world's known plants today. Far from subscribing to a romantic view of photosynthesis such as that put forward by Natasha Myers, as an "utterly magical, totally cosmic alchemical process that tethers earthly plant life in reverent, rhythmic attention to the earth's solar source", from the perspective of plant biology and crop science, photosynthesis is more likely to be described simply as inefficient or even overtly wasteful.[76] Researchers based at the University of Illinois and the Agricultural Research Service of the US Department of Agriculture, for example, have recently sought to boost crop productivity by addressing what they refer to as a "photosynthetic glitch"—the tendency for most plants to absorb not only carbon dioxide but also some oxygen from the atmosphere, thereby generating a need for photorespiration as well as photosynthesis itself.[77] By using genetic transformation techniques to enhance the efficiency of photorespiration, researchers working on this project claim to have freed up more energy for plants to concentrate on the core job of photosynthesis, resulting in a 40 per cent increase in plant biomass achieved by tobacco crops in field trials.[78] Revealingly, in highlighting the potential for photosynthesis itself to be improved through the identification of metabolic shortcuts, such projects often invoke overtly economic metaphors to convey a specific imaginary of what crop plants of the future should look like. Photorespiration, for example, has been described by researchers as a process which "costs the plant precious energy and resources that it could have invested in photosynthesis to produce more growth and yield".[79] From this vantage point, the application of plant biotechnology within American crop science serves—I argue—not strictly to resolve "basic science challenges", as the DOE puts it. Instead, it constitutes an ambitious "energopolitical" project within which plants themselves are conceived as investors whose contributions to future American prosperity hinge on their ability to minimise waste—and maximise their own future growth—by using available energy and resources as efficiently and productively as possible.[80]

In the field of advanced bioenergy research, this use of economic analogies is extended further still. At the Center for Advanced Bioenergy and Bioproducts Innovation (CABBI), for example—the newest of the DOE's four BRCs—plants are conceived not as investors, but rather "as factories ... capable of making fuels and high-value chemical additives directly in their bodies".[81] While researchers at CABBI work on an array of topics, including tackling food and energy security by improving what might be thought of as the raw efficiency of photosynthesis,[82] CABBI is most renowned for its so-called "plants-as-factories" paradigm. Under this approach, scientists employ genetic engineering to radically boost the volume of oil which accumulates not just in the seeds of crop plants, but also directly in their stems and leaves. In so doing, this research "circumvents the challenges of developing efficient lignocellulose deconstruction methods"—challenges which have been central to much of the work undertaken by CABBI's three "sister BRCs" since 2007.[83] Indeed, CABBI's relatively recent emergence in 2017 as a fourth DOE-sponsored BRC might be interpreted in itself as evidence of an ongoing paradigm shift in American advanced bioenergy research. In broad terms at least, this paradigm is not at all new; as early as 1995, the journal *Science* ran a news article headlined "Plants as Chemical Factories", informing readers that biotechnology would soon deliver plants capable of providing "a wide array of nonfood, nonfeed materials", including "biodegradable plastics, industrial lubricants ... as well as drugs and pharmaceuticals".[84] Writing in 2013, meanwhile, sociologist Kenneth Fish had already identified the capacity of biotechnology to mobilise "living factories"—understood not merely as organisms altered through genetic engineering, but as the basis for a "transgenic labour process" in which "the organism *is* the instrument of production".[85] Nonetheless, within the realm of advanced bioenergy research specifically, the idea of plants-as-factories, and the prospect of using this approach to drastically enhance the volume of oil accumulating directly in the biomass of already high-yielding, widely grown agricultural crops, have been described as "ground-breaking".[86]

Whether ground-breaking or not, perhaps the most jarring feature of the bioenergy feedstock development research being undertaken at CABBI is its focus on plant species which have conventionally been cultivated not for oil at all, but rather for starch and sugar. Key species being subject to genetic engineering and field trials under the plants-as-factories paradigm are all herbaceous grass crops—*Miscanthus*, sweet varieties of sorghum, and even sugarcane—rather than dedicated oilseed crops like palm oil or soy. To some extent, this focus can be explained by the already firm reputation of species like *Miscanthus* and sugarcane to deliver high yields; the latter is already the world's most prolific industrial crop measured by total volumes of harvested biomass.[87] By enhancing the proportion of crop biomass that is made up not of lignocellulose, but rather of "ready-made", energy-dense oils known as triacylglycerols (TAGs), researchers hope to provide American farmers not only with grass crops whose oil content significantly exceeds that of conventional oilseed crops like soy, but also with crops from which both sugars and oils can simply be "squeezed out" in the field (see Figure 4.2).[88] Even in some of the earliest DOE-funded research aiming to enhance oil accumulation in sugarcane, predating the formation of CABBI itself, scientists claimed that metabolic engineering could achieve an almost 100-fold increase in the volume of TAGs contained in the crop's stems and leaves.[89] These findings prompted favourable follow-up assessments of the techno-economic feasibility of producing both oil-based jet fuel *and* sugar-based bioethanol from the same sugarcane crop; so-called "oilcane", according to this analysis, "could produce as high as 16 times the jet fuel ... per unit land than that of other oil crops and do so using low-value land unsuited to most other crops, while being highly water and nitrogen use efficient".[90] More up-to-date research, focusing on the development of "oil-rich sorghum", arguably offers less grandiose claims about yield potentials—only a "1.4 times" increase in oil production per hectare over soybeans is mooted, rather than a 16-fold one.[91] Nonetheless, the close entanglement of these

Figure 4.2 A researcher feeds harvested stalks of an experimental sugarcane crop into a three-roller mill at a facility in Louisiana. The mill extracts a dark juice-like cocktail of both sugars and oils.
Source: US Department of Agriculture.

research agendas with policy commitments to net zero aviation in the US remains undeniable; indeed, the ability of TAG oils to help "meet the growing demand for sustainable aviation fuel" is highlighted in the very first sentence of the oil-rich sorghum study, a claim repeated in a CABBI press release of early 2025.[92]

For its scientific protagonists, the plants-as-factories paradigm promises to unlock a utopian future in which molecular-scale carbon allocation and metabolism processes within crop plants are artificially modified to mine the sun directly, thereby furnishing the American bioeconomy with a range of high-value, ready-made bioproducts, whether TAG oils for use as renewable fuel, or other chemical additives for use in manufacturing or medicine. Efforts to develop a bioeconomy are often associated with the concept of "biomimicry", in which nature is supposedly reimagined "as a mentor to be partnered with and revered".[93] The plants-as-factories paradigm, however, can be better understood, I argue, as a kind of "fossil-mimicry" imposed directly onto the world of agriculture. As CABBI's own researchers put it, "[I]t is almost impossible to count the number of goods made from oil and its distilled byproducts. To realize a bio-based economy, plants must do the same thing: make many useful products from a single source".[94] In upholding oil as a blueprint to be emulated, such visions invoke the idea not merely of plants-as-factories, but, more specifically still, of "plants-as-refineries". In both cases, moreover, the implication is that agricultural crops should be understood not just in terms of the efficiency and productivity of their energy and resource use as individual plants, but more radically as industrial facilities—or indeed workplaces—within which the mass production and assembly of high-value commodities can be orchestrated and flexibly reconfigured at will. Carbon allocation and metabolism within agricultural crop plants, from this standpoint, constitute customisable processes with the potential to be reordered and remade in theoretically limitless ways through genetic engineering. Recalling former President Biden's bold claim about "the potential of biology to make *almost anything* that we use in our day-to-day lives",[95] this is then a paradigm which apprehends plant biology, following Melinda Cooper and Catherine Waldby, as "a source of flexible, non-exclusive potentialities, any one of which can be materialized in practice".[96] By foregrounding the manipulability of biological processes, however, this is also a paradigm which contains within itself the seeds of a potential critique of the productivist logics which, through discursive recourse to oil, are presently being upheld as inherently desirable. What to do with plants on a molecular level, how to make them "better" and for which societal or economic ends are not scientific questions but normative ones—and hence questions to which the tools of biotechnology alone can offer no automatic or pregiven correct answers.

Living oil fields: hiding vegetal labour in plain sight

Appearing on BBC Radio 4's *Today* programme on 30 January 2025, UK Chancellor of the Exchequer Rachel Reeves, pressed to explain her government's endorsement of plans to construct a third runway at London's Heathrow Airport, described SAF as "a game changer in the way that we fly", before adding, "[W]e know that sustainable aviation and economic growth go hand-in-hand".[97] Despite her optimistic outlook, Reeves offered no specific details about the kinds of resources—whether waste-based, forest-based, crop-based or otherwise—that would need to be harnessed to actually produce large volumes of SAF. This omission came despite the Royal Society having published a policy brief less than two years previously, finding that to replace all UK aviation fuel with biofuels would require over half of the country's existing agricultural land.[98] As advanced biofuels increasingly come to be seen as critical to the future of an industry which propels aircraft through rarified skies some five miles or more above sea level, there is a certain irony to be found in politicians speaking of SAF as if it can be made to emerge not from farms or forests, but rather from thin air. Reeves's elision of the supply-side realities of SAF production, however, was not unusual. Instead, it serves as just one example of a much wider trend in policy discourse around advanced biofuels, of downplaying or even simply ignoring the complex material and spatial requirements of large-scale bioenergy resource-making in practice.

This chapter has offered a detailed analysis of unfolding efforts to bring into being a vast new advanced bioenergy resource in the US, one that might be capable not merely—as if this objective alone were unambitious—of enabling the country's entire aviation sector to go net zero by the middle of this century, but also of furnishing the American bioeconomy with an abundance of other sustainable biofuels and high-value bioproducts besides. On one level, the chapter has shown—focusing on crop-based biofuels specifically—how official US government reports on the potential scalability of the nation's sustainable biomass supply are undergirded by specific, narrow forms of expert knowledge, calculation and representation. By mobilising concepts and techniques from agricultural economics and land-use modelling, and by using cartography to offer compelling "statistical pictures" of an as-yet-unrealised American bioenergy resource, these reports strip farming of its rich social and cultural meanings, reducing landowners' planting and management decisions to predictable, rational responses to changing price signals and evolving market demand. Agricultural land itself, meanwhile, is represented in maps as at once a unified object and also a passive one, appearing as a vast repository of fields and soils awaiting improvement and redemption through the arrival of more resilient and productive crop cultivars, including those tailored specifically for enhanced oil accumulation. On a second level, meanwhile, the chapter has also scrutinised US government-funded scientific engagements with plant life in fields concerned with the development of more advanced bioenergy feedstocks, arguing that these engagements go far beyond seeking to resolve "basic scientific challenges". Instead, research in this vein—I have argued—weaves plant biology, crop science, molecular genetics and biotechnology together into an explicitly energopolitical project, one which regards current forms of plant metabolism as inherently inefficient and wasteful, thereby requiring remediation in order to properly unlock the full fuel-giving potential of agriculture itself. Plants, from this perspective, should not be understood strictly as individual agents of production or growth, but rather as factories—that is, as industrial facilities within which the mass production and assembly of high-value commodities, including but not limited to combustible oil, can be staged and reconfigured at will.

At the levels of both official policy reports and techno-economic analyses, and in the ontological approach to plants exhibited by US government-funded bioenergy research programs, the hypothetical enrolment of vast amounts of future plant photosynthesis into large-scale modern bioenergy projects—whether to realise a net zero aviation future or to achieve other goals—is thus treated a priori as both a feasible and a desirable objective. A view of the dedicated energy crops of the future as factories, in other words, doesn't just take for granted the idea that plant metabolism should be geared towards facilitating the mass production of renewable fuels and other high-value commodities. It also elides the existence of alternative possibilities for enrolling plant growth into more diverse human social or economic projects. In concluding this chapter, I want to suggest that to help resurface these possibilities, we would do well to return to Kenneth Fish's insistence that the creation of living factories entails not simply the modification of organisms on a molecular level, but more fundamentally the setting up of a "transgenic *labour* process".[99] To conceive of genetically engineered crop plants as sources of labour power rather than as factories, I argue, is to insist upon the contingency and contestability of the end goals for which those plants' metabolism—in the form of photosynthesis, carbon accumulation and biomass growth—is harnessed. Indeed, rather than seeing crop plants as passive entities to be manipulated at will, a view of those same plants as workers would force us—as I have argued in earlier chapters—to recognise that all bioenergy resource-making, whether driven by visions of a net zero aviation future or alternatively by a desire to meet basic human requirements for warmth and cooking fuel, is inherently reliant upon the "hybrid labour" of humans and plants working together.[100]

The urgency of exposing these contingencies, and of foregrounding the *work* that plants do in producing biofuels of all kinds—whether supposedly advanced or otherwise—can be located precisely in the danger that modern bioenergy never comes to be recognised as a contestable idea, rather than as a technical solution to be marshalled as part of a broader, so-called clean energy "transition". By recognising plants as labourers, we give ourselves the opportunity to rethink how we might work with plants not just to maintain but potentially also to reshape the economic and social systems which have been created and sustained by the large-scale burning of fossil hydrocarbons. To return to grandiose visions of a net zero aviation sector underpinned by massive volumes of SAF, asking such questions might serve to foreground questions not just about aircraft fuel but also about flying's relationship to economic growth and human wellbeing, for example. As things stand, however, the paradox of SAF—as with advanced bioenergy more broadly—is that it ostensibly offers a "transition" away from fossil fuels, even as the slow, faltering, never-quite-ready to be scaled-up status of these kinds of biofuels serves in practice to shore up the ongoing political and cultural legitimacy of incumbent fossil fuel-based industries and infrastructures. Far better, one might argue, to wrestle now with the difficult issue of what kinds of economies and societies we would like human–plant relationships to help fuel in the coming decades, than to leave those questions hidden behind a blind faith in the ability of a wholesale transition to modern bioenergy—a transition which may never arrive—to finally resolve the contradictions between intensive hydrocarbon burning and a warming planet. Questions about how we might work otherwise with plants, and what plants might have to teach us about the relationships between energy, productivity, growth and human progress, will be returned to in more detail in the book's concluding chapter. Beforehand, however, the book first turns its attention to a final case study of modern bioenergy as a regulatory logic—that of BECCS and its association with visions of large-scale hydrocarbon burning not simply as a GHG emissions problem to be mitigated, but rather as a potential tool of achieving negative emissions and, hence, climate repair.

Notes

1 BBC News, "Groundbreaking transatlantic flight using greener fuel lands in the US."
2 Virgin Atlantic, "Flight 100."
3 BBC News, "Groundbreaking transatlantic flight using greener fuel lands in the US."
4 International Air Transport Association, "Net-zero carbon emissions by 2050."

5 Clean Skies for Tomorrow, "2030 ambition statement." Some national governments have announced similar targets; the UK aims to meet a target of at least 10 per cent of SAF in its "aviation fuel mix" by 2030 (see UK Department for Transport, *Jet Zero Strategy*, 9).
6 Virgin Atlantic, "Flight 100."
7 Mission Possible Partnership, *Making Net-Zero Aviation Possible.*
8 Mission Possible Partnership, *Making Net-Zero Aviation Possible*, 57.
9 International Air Transport Association, *Net Zero 2050*, 2.
10 Mission Possible Partnership, *Making Net-Zero Aviation Possible*, 57.
11 CABBI, *Filling the Gaps and Defining the Future of Bioenergy.*
12 US Department of Energy *et al.*, *Memorandum of Understanding: Sustainable Aviation Fuel Grand Challenge*, 1.
13 US Department of Energy *et al.*, *SAF Grand Challenge Roadmap*, 1.
14 US Federal Aviation Administration, *United States 2021 Aviation Climate Action Plan.*
15 National Renewable Energy Laboratory, *Sustainable Aviation Fuel (SAF) State-of-Industry Report*, 2.
16 US Department of Energy *et al.*, *SAF Grand Challenge Roadmap*, 9.
17 US Department of Energy *et al.*, *SAF Grand Challenge Roadmap*, 16.
18 The White House, "Remarks by President Biden on helping workers and innovators invent and make more in America."
19 US Department of Energy *et al.*, *Memorandum of Understanding: Sustainable Aviation Fuel Grand Challenge*, 1.
20 US Department of Energy *et al.*, *SAF Grand Challenge Roadmap*, vii (emphasis added).
21 US Department of Energy *et al.*, *SAF Grand Challenge Roadmap*, 12.
22 US Department of Energy *et al.*, *SAF Grand Challenge Roadmap*, 12–13.
23 Birch and Calvert, "Rethinking 'drop-in' biofuels," 54.
24 Tyner, "Why the push for drop-in biofuels?"
25 Fougeres, "First solely-biofuel jet flight raises clean travel hopes."
26 US Department of Energy *et al.*, *SAF Grand Challenge Roadmap*, 65; and Renshaw, Kelly, and Douglas, "Biden administration split over ethanol's role in aviation fuel subsidy program."
27 Bearak, Searcey, and Rojanasakul, "Airlines race toward a future for powering their jets with corn."
28 Reuters, "Biden team sets out path for ethanol aviation fuel subsidies."
29 The White House, "President signs Energy Policy Act"; see also Chapter 2.
30 The White House, "Executive order on advancing biotechnology and biomanufacturing innovation for a sustainable, safe, and secure American bioeconomy"; see also Asdal *et al.*, " 'The good economy'."
31 The White House, "Fact sheet: President Biden to launch a National Biotechnology and Biomanufacturing Initiative" (emphasis added).
32 The White House, *Bold Goals for US Biotechnology and Biomanufacturing.*
33 US Department of Energy *et al.*, *SAF Grand Challenge Roadmap*, 17.
34 Parajuli *et al.*, "Towards oilcane."

35 US Department of Energy *et al.*, *SAF Grand Challenge Roadmap*, viii.
36 US Department of Energy, *2016 Billion-Ton Report*.
37 The White House, "President Bush delivers State of the Union address."
38 US Department of Energy, *2023 Billion-Ton Report*.
39 US Department of Energy, *2023 Billion-Ton Report*, xxi.
40 US Department of Energy, *2023 Billion-Ton Report*, xxi.
41 US Department of Energy, *2023 Billion-Ton Report*, 7.
42 US Department of Energy, *2016 Billion-Ton Report*, 142.
43 US Department of Energy, *2023 Billion-Ton Report*, xxi.
44 Oomen *et al.*, "Techniques of futuring," 258.
45 Beck and Mahony, "The politics of anticipation," 2.
46 US Department of Energy, *2023 Billion-Ton Report*, xx. Considered separately from the various "mature-market" scenarios is a further set of modelled circumstances in which long-term biomass demand is deemed sufficiently strong to propel "future innovations" in the production of biomass from micro- and macroalgae, leading to significant additional supply from these sources.
47 US Department of Energy, *2023 Billion-Ton Report*, 70.
48 US Department of Energy, *2023 Billion-Ton Report*, 43.
49 US Department of Energy, *2023 Billion-Ton Report*, 109.
50 US Department of Energy, *2023 Billion-Ton Report*, 19.
51 US Department of Energy, *2023 Billion-Ton Report*, 113.
52 US Department of Energy, *2023 Billion-Ton Report*, 101.
53 US Department of Energy, *2023 Billion-Ton Report*, 104.
54 US Department of Energy, *2023 Billion-Ton Report*, 103.
55 US Department of Energy, *2023 Billion-Ton Report*, 121.
56 US Department of Energy, *2023 Billion-Ton Report*, 264.
57 See the US Department of Energy's "Bioenergy KDF [Knowledge Discovery Framework]" at https://bioenergykdf.ornl.gov.
58 Bridge, "Resource geographies II," 120.
59 Beck and Mahony, "The politics of anticipation," 2.
60 Scott, *Seeing Like a State*, 11.
61 In its treatment of the potential scale of sustainable biomass supply from American agriculture, for example, the *2023 Billion-Ton Report* admits that it "does not make any explicit assumptions on the impacts of climate change in US or international crop yields" (US Department of Energy, *2023 Billion-Ton Report*, 103).
62 See van der Horst and Evans, "Carbon claims and energy landscapes," 190.
63 See Boyer *et al.*, "The making of sustainability," 682–683.
64 Demeritt, "Scientific forest conservation and the statistical picturing of nature's limits in the Progressive-era United States."
65 Demeritt, "Scientific forest conservation and the statistical picturing of nature's limits in the Progressive-era United States," 444, 431; see also Mitchell, *Rule of Experts*.
66 Demeritt, "Scientific forest conservation and the statistical picturing of nature's limits in the Progressive-era United States," 431.

67 Demeritt, "Scientific forest conservation and the statistical picturing of nature's limits in the Progressive-era United States," 433.
68 Goldstein, "*Terra economica*," 357.
69 This section's heading borrows from the title of Deborah Fitzgerald's 2010 book on the industrialisation of American agriculture, *Every Farm a Factory*.
70 US Department of Energy, "Bioenergy Research Centers."
71 US Department of Energy, "DOE announces $590 million to increase bioenergy research."
72 US Department of Energy, "U.S. Department of Energy announces $61 million for biofuels research to reduce transportation emissions."
73 Zoghlami and Paës, "Lignocellulosic biomass"; and Bakker, "Neoliberal nature, ecological fixes, and the pitfalls of comparative research."
74 Science News, "A plant for power alcohol," 10a; see also Manuel, "Lessons from a forgotten fuel."
75 Padella *et al.*, "What is still limiting the deployment of cellulosic ethanol?" 1.
76 Myers, "Photosynthesis."
77 Realizing Increased Photosynthetic Efficiency, "Scientists engineer shortcut for photosynthetic glitch, boost crop growth by 40 percent."
78 South *et al.*, "Synthetic glycolate metabolism pathways stimulate crop growth and productivity in the field."
79 Realizing Increased Photosynthetic Efficiency, "Scientists engineer shortcut for photosynthetic glitch, boost crop growth by 40 percent."
80 See Boyer, "Energopolitics and the anthropology of energy"; and Daggett, *The Birth of Energy*.
81 CABBI, *Filling the Gaps and Defining the Future of Bioenergy*, 4.
82 CABBI, "Rubisco's role in global food and energy security."
83 CABBI, "Feedstock production."
84 Moffat, "Plants as chemical factories," 659.
85 Fish, *Living Factories*, 53.
86 CABBI, "Building a better bioenergy crop."
87 Henry, *Plant Resources*, 69.
88 CABBI, *Filling the Gaps and Defining the Future of Bioenergy*, 4.
89 Zale *et al.*, "Metabolic engineering of sugarcane to accumulate energy-dense triacylglycerols in vegetative biomass"; see also Parajuli *et al.*, "Towards oilcane."
90 Kumar *et al.*, "Biorefinery for combined production of jet fuel and ethanol from lipid-producing sugarcane," 92.
91 Park *et al.*, "Development of vegetative oil sorghum."
92 Park *et al.*, "Development of vegetative oil sorghum," 660; and CABBI, "From lab to field."
93 Goldstein and Johnson, "Biomimicry," 75.
94 CABBI, *Filling the Gaps and Defining the Future of Bioenergy*, 4.
95 The White House, "Fact sheet: President Biden to launch a National Biotechnology and Biomanufacturing Initiative" (emphasis added).

96 Cooper and Waldby, *Clinical Labor*, 13.
97 Georgiadis and Pickard, "UK forecasts small role for green airline fuel by 2040."
98 Royal Society, *Net Zero Aviation Fuels*, 22.
99 Fish, *Living Factories*, 53 (emphasis added).
100 Battistoni, "Bringing in the work of nature"; and Palmer, "Putting forests to work?"

5

Going global: mobilising vegetal labour for climate stabilisation

> Even very low [emissions] limits are eminently feasible. You just have to give yourself the freedom to first go beyond them.
>
> —Andreas Malm and Wim Carton, *Overshoot*, 2024

In February 2019, Drax power station—the UK's largest piece of electricity-generating infrastructure—became the first such facility in the world to capture carbon dioxide emissions generated exclusively from the burning of biomass fuel. As part of a six-month pilot demonstration project, carbon that would normally have escaped into the atmosphere was instead extracted from post-combustion flue gases using a solvent-based technology developed by C-Capture—a spin-out company associated with the School of Chemistry at the nearby University of Leeds.[1] This was by no means the first attempt—either in the UK or elsewhere—to capture emissions of GHGs produced during large-scale electricity generation. Significant hopes have long been pinned on carbon capture and storage (CCS) by scientists, industry groups and policymakers seeking to curb the climate impacts of the countless fossil fuel power plants upon which global society continues to depend heavily for its day-to-day energy needs.[2] Whereas previous attempts to retrofit CCS technology to power plants had, however, been concerned with keeping long-buried, subterranean reserves of carbon—contained principally in coal and natural gas—out of the atmosphere, Drax's demonstration project was engaged in a far more ambitious endeavour. After all, the carbon dioxide being released here had originally been sequestered not millions of years ago, but much more recently, by trees growing in intensively managed forests in places like the US South, Canada, Finland and other European countries.[3] The processing of these trees into a combustible hydrocarbon fuel—specifically, compressed wood pellets—was an even more recent achievement still. Successfully capturing carbon from the flue gases produced by burning these wood pellets could therefore be construed as a feat not simply of reducing emissions, but of actively removing carbon dioxide from the atmosphere itself. In short, biomass-based electricity generation, if only for a brief moment, had been rendered into a carbon-negative technology.

While unprecedented in many respects, this pilot demonstration of BECCS was nonetheless modest in scale. Only approximately one tonne of carbon dioxide per day was prevented from escaping into the atmosphere during the trial, and even then only temporarily. Once it had been successfully separated out from the other flue gases being generated at the plant, the captured carbon dioxide was simply re-released, rather than being permanently stored. Set against the more than 15 million tonnes of carbon dioxide that are released annually by burning biomass fuels to generate electricity in the UK, the capture of just a single tonne per day might be regarded as paltry.[4] One representative of C-Capture conceded as much about the trial in acknowledging that "the challenge now is to get all the information we need to design and build a capture plant 10,000 times bigger".[5] Yet it was precisely by offering up the tantalising prospect that biomass fuels might one day serve as the basis not only for large-scale renewable electricity generation, but also for achieving massive quantities of negative GHG emissions, that this achievement was greeted as a major milestone not just by Drax but also by the broader UK climate policy community. Indeed, the scaling up of BECCS is widely viewed—by a range of influential advisory and academic bodies in the UK, including the Committee on Climate Change, the Royal Society and the Royal Academy of Engineering—as an indispensable component of the country's efforts to reach an overall GHG emissions balance of net zero by the middle of this century.[6] Certainly there is some evidence that the deployment of BECCS is increasing globally—a recent study estimates that around 0.5 million tonnes of carbon was removed from the atmosphere using this method in 2023.[7] Nonetheless, according to one prominent report, the UK alone would need to achieve around one hundred times this much BECCS every year—50 million tonnes of carbon removals annually—to be assured of meeting its own target of net zero by 2050.[8] On a global scale, meanwhile, the Sixth Assessment Report of the IPCC, published in 2022, suggests that limiting climate warming to 2 degrees Celsius will require removing a cumulative total of between 170 and 650 *billion* tonnes of carbon dioxide from the atmosphere using BECCS by the year 2100.[9]

This chapter will examine how an ongoing clamour for BECCS to be rapidly expanded, ostensibly essential if global temperature rises above pre-industrial levels are to be restricted to no more than 2 or even 1.5 degrees,[10] is profoundly reshaping contemporary energy resource-making practices and economies. It begins by scoping out the history of the idea of combining bioenergy with CCS to realise negative GHG emissions, following its evolution in international scientific and policymaking debates from a fringe concept to a now fundamental component of almost all projected future global GHG emissions scenarios. Having charted the rise of BECCS as a discursive idea, the chapter will then move on to scrutinise the technology's more tangible implications for both the shifting geography and the evolving materiality of twenty-first-century energy systems. While these implications are certainly severe, the chapter will insist that to expose the full political stakes of BECCS requires paying closest attention of all to the novel relationships the technology is serving to establish between vegetal life, energy consumption and new kinds of economic value. More specifically here, the chapter will contend that BECCS is noteworthy not just because it promises to deliver large-scale negative GHG emissions, but also because of the role that it envisages for plants and trees as motors of novel forms of value generation, importantly predicated not on reducing emissions, but on harnessing carbon dioxide as a means of production to be put to use in its own right. Indeed, whether through plans to produce sustainable concrete, proteins for animal and fish feed, bioplastics or even synthetic fuels, carbon dioxide—as captured by vegetal lifeforms—is being rapidly reappraised by proponents of BECCS as a raw material with diverse economic potentials of its own. As a result, the chapter will contend, plant growth is being reenvisaged today as an engine of new forms of wasteless metabolism, in a manner which promises only to deepen the interconnections between large-scale energy consumption, economic growth and wider societal notions of human progress. The real political stakes of BECCS, in short, can be found in the claim which this technology lays—on a global scale—to plant life itself, understood as a theoretically inexhaustible basis not just for reinvigorating industrial economies, but for solving the climate change crisis too.

A brief history of BECCS: game changer or dodgy fix?

Scientific work on BECCS can be traced back to the late 1990s and early 2000s, even if the term itself was not coined until 2003.[11] Initial work on the idea was linked to the potential for industrial actors already making use of biomass resources to gain economic advantages through emergent GHG emissions trading markets, as provided for under (what was then) the recently ratified Kyoto Protocol. One early prototypical form of BECCS, for instance—floated by Kenneth Möllersten of the Royal Institute of Technology in Stockholm—was pitched as a means of enabling Swedish paper and pulp mills to earn not just one, but two, sets of carbon credits in such markets.[12] The potential significance of these proposals for broader, international efforts to mitigate climate change was readily apparent, however, including for researchers such as Michael Obersteiner, then based at the International Institute for Applied Systems Analysis (IIASA). In mid-2002, Obersteiner, Möllersten and several colleagues co-authored an IIASA report examining "the properties of biomass-based sequestration technologies" and "their potential role in climate risk management".[13] The report acknowledged that what it termed "Biomass Energy Carbon Sequestration (BECS)" was "in its primordial stage" and would be likely to raise a number of environmental and social sustainability issues if ever deployed at scale.[14] Yet, placing heavy emphasis on the "volatile and highly uncertain" nature of the global climate system's future response to steadily increasing concentrations of GHG, the report also contended that BECS—like other negative emissions technologies—should be recognised and seriously considered as an "important 'real option' nested in a wider portfolio of climate mitigation technologies".[15]

Especially notable in the IIASA report was a view of the planet's "natural" carbon sinks—whether in the form of terrestrial ecosystems such as forests and peatlands, or indeed marine ecosystems—as inherently limited in their ability to compensate for rising anthropogenic emissions of carbon dioxide. The potential for these carbon sinks to take on a more intensive role in sequestering carbon, for example, was deemed likely to be compromised not only by the natural tendency of organic matter to decay over time, but also by the prospect that forests and other land-based ecosystems might themselves be destabilised by climate change in the relatively near future. Given the "rather limited capacity" of these sinks—and even then, a capacity of "uncertain permanence"—Obersteiner and colleagues cautioned strongly against pinning hopes of successful climate change mitigation on any significant enhancement to future rates of ecosystem-based carbon sequestration.[16] The appeal of BECS, by contrast, could be found precisely in its ability to significantly increase planetary carbon sink capacity without having to rely on the storage potential of terrestrial or marine ecosystems. In this view, the carbon sinks constituted by living ecosystems could be actively complemented, using BECS, by engineered repositories for the long-term storage of carbon, including potentially in some of the same oil and gas fields whose depletion had brought about climate change in the first place. In other words, by "scrubbing" CO_2 directly from the flue gases released during biomass combustion, the Achilles' heel of so-called "natural" terrestrial carbon sinks—the inevitable decay of organic matter—could be intercepted and harnessed in a controlled manner, as both an energy resource and a means of removing carbon dioxide permanently from the atmosphere. In this way, BECS would divert pressure away from the earth's "natural" biological carbon sinks, and towards an alternative, entirely anthropogenic one.

While the IIASA report is largely preoccupied with assessing the technical feasibility of achieving meaningful amounts of GHG emissions removal using BECS, it nonetheless paved the way for a radical reappraisal of the roles that vegetal lifeforms might play in helping to counteract and mediate the effects of anthropogenic climate change. Indeed, in contrasting BECS directly with existing terrestrial and marine-based carbon sinks, the report envisions a response to climate change in which plants are increasingly relieved of the burden of keeping carbon dioxide locked up in their own biomass. Instead, their role in climate mitigation is now precisely to be burned in industrial facilities where the resulting emissions of carbon dioxide can be captured and ultimately stored in human-made repositories. As demonstrated by this book's earlier analysis of bioenergy resource-making practices in working forests of the US South (see Chapter 3), such a view of plants and trees is today increasingly standardised within modern bioenergy discourses. In the early 2000s, however, these propositions were not merely innovative on their own terms; they were also understood to have distinctly radical implications for the wider landscape of climate change policy and action. It is not just that what came to be known as BECCS was understood as a means of radically reconfiguring flows and cycles of carbon at the planetary scale, thereby augmenting the earth's capacity to absorb the waste products associated with intensive hydrocarbon consumption.[17] Just as importantly, BECCS was also pitched from the outset as a potential catalyst for bringing into being more ambitious overall targets for reducing carbon dioxide concentrations in the atmosphere, not least to guard against the risk that the climate system might ultimately prove more susceptible to GHG forcing than had previously been expected. As the IIASA report put it, a "big part of the value" of combining bioenergy with CCS, as a negative emissions technology, was—initially at least—to be found in this technology's ability to "make it possible to meet *more stringent* CO_2 concentrations than many people today think is possible".[18]

Despite these political potentials, however, the extent to which BECCS has subsequently been associated with the formulation of more ambitious or indeed more stringent climate change targets, at least at the global scale, is highly questionable. Certainly, interest in the potential of BECCS—alongside other negative emissions technologies—has coincided with the emergence, since the mid-2010s, of an overarching new tier of aspiration to restrict global warming to no more than 1.5 degrees above preindustrial levels, represented most clearly by the 2015 Paris Agreement and a subsequent IPCC special report.[19] Hiding behind these headline (and, importantly, legally non-binding) commitments, however, BECCS has also rapidly become a mainstay of future global GHG emissions scenarios compatible with meeting the much longer-standing—and almost universally accepted—"maximum allowable warming" target of 2 degrees.[20] In fact, having only been formally acknowledged by the IPCC for the first time in 2007, by 2014, more than 100 of the 116 scenarios set out by the organisation in its Fifth Assessment Report on Climate Change, all intended to meet this well-worn 2 degree target, relied on the deployment of some level of negative emissions technologies, with BECCS shouldering a large degree of the burden.[21] In these scenarios, moreover, BECCS serves typically either to compensate for so-called "residual emissions" in sectors where GHG emissions reductions are expected to be especially difficult to achieve, such as aviation and agriculture, or otherwise to "return [global] emissions to target levels after an overshoot resulting from a lack of near-term action".[22] That the IPCC's climate change mitigation pathways have come to rely "almost completely" on negative emissions technologies—with BECCS playing the most prominent role—exposes net zero discourse itself as a means of "reducing the planetary bio-geophysical system to a ledger of debits and credits".[23] Moreover, this dependency on BECCS has become necessary simply in order for models to continue to demonstrate the feasibility of meeting global temperature targets which, less than two decades previously, had been deemed achievable through conventional GHG emissions reduction measures alone.[24] For Michael Obersteiner, commenting in 2016, BECCS has thus not been deployed, as was originally intended, as a tool for enabling genuinely "ambitious climate targets"—instead, it has been "unfortunately misused for regular scenarios".[25]

The idea of BECCS today is therefore arguably a far more conservative proposition than the one which initially entered global climate policy networks in the early 2000s. To an extent, of course, one might contend that this is not surprising—concentrations of carbon dioxide and other GHGs in the atmosphere have risen precipitously over the past two decades, rapidly increasing the severity of the societal responses that will be required to constrain average surface temperature rises to no more than 2 degrees above preindustrial levels (if such a target is to be met at all). Accordingly, BECCS has come to occupy a central position not just in the world of climate modelling (featuring in 95 per cent of the more than six hundred distinct GHG emissions scenarios surveyed by one recent landmark study)—it also increasingly attracts significant grant funding as well (see Figure 5.1). Despite these shifts, however, limited global deployment means that the technology remains "highly speculative".[26] Moreover, there is a dearth of relevant governance frameworks, either internationally or otherwise, intended explicitly to incentivise its implementation over the coming decades.[27]

A large part of the explanation for this mismatch can be found in the logic of the integrated assessment models (IAMs) which undergird the

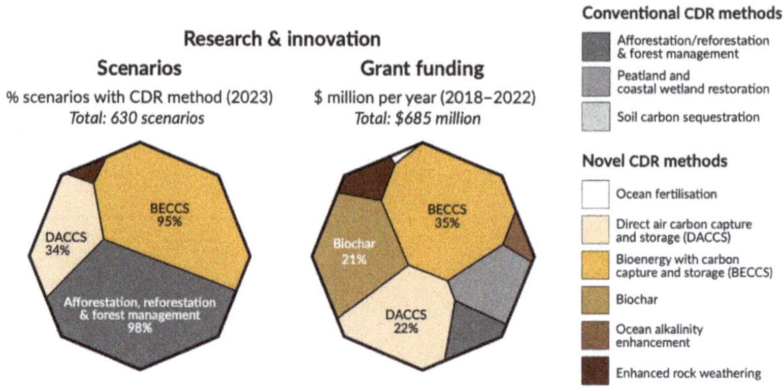

Figure 5.1 The prominence of BECCS in a suite of 630 published GHG emissions scenarios and as a recipient of grant funding (2018–22), as compared with eight other methods for removing carbon dioxide from the atmosphere.
Source: Smith *et al.*, *The State of Carbon Dioxide Removal 2024*, 2nd ed., 13.

IPCC's mitigation pathways. As geographer Wim Carton has highlighted, IAMs are ultimately centred around "the prioritisation of cost effectiveness as a seemingly neutral method to arbitrate between alternative climate futures".[28] Given that cost optimisation models of this kind typically apply a "discount" rate of around 5 per cent, forms of climate action taken in the future—including the large scale-deployment of BECCS—inevitably appear less expensive, in relative terms, than the "more radical short term mitigation" actions that would otherwise be required to meet a 2 (or indeed 1.5) degree target.[29] Even as it aims to assist policymakers in preparing "neutral" assessments, discounting can of course be viewed as an explicitly "political technology"—not least because, in the context of BECCS, it has already helped to postpone the point at which the use of fossil fuels to generate energy will become unprofitable in practice.[30] In other words, the mere invocation of large-scale BECCS deployment in the future, however promissory or speculative, is already having significant "world-making" consequences in the present, acting to disguise a "political economy of delay" behind the facade of ostensibly politically neutral commitments to cost optimisation.[31]

In light of these dynamics, geographers and political ecologists have begun to raise probing questions about the kind of response to climate change that BECCS ultimately represents. At one level, BECCS appears to conform neatly to the idea of a "socioecological fix" for climate change, promising as it does to provide "opportunities for the reinvigoration of capital accumulation on a global scale and a biophysically significant response" to rising global temperatures.[32] Yet the incontrovertibly promissory nature of BECCS—and especially the fact that it enables long-term climate mitigation targets to be depicted as feasible even as GHG emissions continue to rise—also invites a more cynical view, wherein the "fix" being advanced is not for climate change itself, but rather for the threat of devaluation faced by existing "carbon-intensive accumulation processes", including those in the fossil fuel sector itself.[33] Much-rehearsed descriptions of BECCS as a "game changer", whether issued on behalf of start-up firms developing solvent-based carbon capture technology in the UK[34] or by institutions representing incumbent industry interests, such as the Energy Technologies Institute,[35] take on a particularly intriguing appearance when viewed against this backdrop. After all, if BECCS does indeed change the rules of an ongoing game in climate change politics, this begs further questions: Who exactly are the various players in that game, and precisely how are the power relations between those players being reconfigured by the prospect of large-scale negative emissions? What is abundantly clear is that BECCS has already, without needing to be implemented on any significant scale, exacerbated and intensified a pre-existing set of contradictions in global climate policy between "talk, decisions, and action".[36] To paraphrase anthropologist Steve Rayner, talking up the promise of BECCS—not least by incorporating those promises into future emissions scenarios endorsed by the IPCC—is no guarantee of bringing about positive results in practice; such expectations, ultimately, are the realm of "magical thinking".[37] Moreover, and as the next section will argue, inconsistencies between talk, decisions and action become even more stark when the material and practical realities of implementing BECCS proposals on the ground are considered in detail.

Large-scale BECCS: climate solution or self-delusion?

If BECCS is to achieve large-scale negative GHG emissions this century, enormous swathes of forests and arable land will inevitably have to be enrolled in the production of the necessary biomass feedstocks. All renewable energy technologies take up some amount of dedicated space on the earth's surface, of course. Indeed, as noted in earlier chapters, for geographers Matt Huber and James McCarthy, we are currently in the midst of an ongoing global transition, from a subterranean energy regime predicated upon fossil fuels to a thoroughly "horizontal" regime predicated on an array of renewable resources available to be harnessed at the earth's surface (whether in the form of incoming solar radiation, winds or indeed tides).[38] As discussed in Chapter 1, the acquisition of land for renewable energy projects—whether for solar parks, wind farms or otherwise—has already been shown to have potentially pernicious social consequences.[39] Yet there are at least two important respects in which the future dedication of large tracts of land to biomass feedstock production for BECCS will be distinct from other aspects of the world's increasingly "horizontal" energy regime. The first is that the cultivation of biomass feedstocks is in practice often virtually indistinguishable, at least to the untrained eye, from forms of agriculture or forestry which have nothing to do with energy provision. As such, it is not always immediately obvious when transformations in the use of land for bioenergy production have taken place, making it difficult to monitor precisely where bioenergy is coming from and how it is being produced in real time. A second and more fundamental difference, however, is simply that through its combination with CCS, bioenergy is alone among existing renewable energy technologies in being expected to help global society achieve negative GHG emissions (as opposed to merely reducing them as close to zero as possible). If these expectations are to be met, the size of the spatial footprint occupied by BECCS during the course of this century will therefore almost certainly exceed that of other renewables by a significant margin.

How much land exactly would need to be given over to the cultivation of biomass feedstocks to make BECCS work? Since the efficiency of feedstock cultivation for bioenergy varies widely, so too do estimated answers to this question. According to one major study of the potential global impacts of a range of negative emissions technologies, published in 2016, between 3.8 million and 7 million square kilometres of land would need to be devoted to biomass feedstock cultivation by the year 2100 if BECCS were to successfully restrict global warming to no more than 2 degrees above preindustrial levels.[40] More recently, the IPCC has suggested that to keep temperatures below 1.5 degrees would require dedicating approximately two million square kilometres of land to energy crop cultivation for BECCS by the much earlier date of 2050.[41] Uncertainties over the efficiency of crop cultivation, however, mean this figure could reasonably be much closer to five million square kilometres in reality. In the vast majority of emissions scenarios, it is presumed that BECCS would be fuelled principally by the cultivation of "dedicated, high productivity energy crops".[42] Assuming that around five million square kilometres of land could need to be given over to this purpose by the middle of this century, it is sobering to reflect, as social scientist Holly Jean Buck has recently done, that this equates to "an area one and a half times the size of India".[43]

It is not just the scale of these implicit land-use commitments which is concerning, however. As was discussed in Chapters 2 and 3, the expansion of modern bioenergy systems which has occurred in many developed economies to date—whether to replace coal burning in the electricity sector, or to replace petrol and diesel in the transport sector—has relied heavily upon imported feedstocks. Any effort to deploy BECCS on the vast scales being envisaged by the IPCC and other influential bodies today would therefore carry a substantial risk of deepening a decidedly neocolonial set of relationships between less affluent biomass-producing countries, generally located in the global South, and more affluent biomass-consuming countries largely concentrated in the global North.[44] Indeed, as countries like South Korea and Japan—among others—follow the US and many European economies in committing to replace more of their fossil energy resources with imported biomass fuels, competition for the trappings of global photosynthetic activity looks set to increase substantially in the coming decades. A 2020 report prepared for the UK Department for Business, Energy and Industrial Strategy, for example, predicts that "UK access to some of the global biomass sources would decrease to 2050 as countries establish their own biomass plants or [through] inter-regional competition".[45] Certainly, if large-scale BECCS were to become not only a mainstay of projected GHG emissions scenarios, as set out in expert scientific reports, but also a legally binding commitment enshrined in the climate change policies of multiple affluent countries, the result could be nothing less than a new scramble for control of the world's remaining under-utilised agricultural land—a scramble in which both the land rights and the future economic autonomy of populations living across much of the global South would be likely to suffer considerably. Merely bolting CCS technology onto the end of any massively expanded global network of bioenergy supply chains, in short, will do nothing by itself to reduce the threat of a renewed wave of land grabbing in these settings, potentially even larger in scope than that associated with the new biofuel rush of the early 2000s.[46]

Equally unaffected by the addition of CCS technology to bioenergy supply chains are the range of fundamental inefficiencies and logistical challenges associated with biofuel production, as already discussed in Chapter 2. Accordingly, the sometimes significant water and fertiliser requirements associated with crop cultivation, along with the large energy inputs necessary to process and transport what are often bulky feedstocks, as well as wider issues associated with direct and indirect land-use change—including potentially significant GHG emissions, biodiversity losses and rising food insecurity—must all be factored into any full assessment of BECCS.[47] One recent analysis has even suggested that some BECCS supply chains could, when these issues are properly taken into account, end up delivering carbon-positive, rather than carbon-negative, outcomes.[48] Capturing and storing carbon dioxide, moreover, will themselves require costly and potentially emissions-intensive investments in new infrastructures, not least vast pipeline networks capable of transporting the gas away from power stations and on to sequestration sites in a safe and efficient manner. For Holly Jean Buck, these various issues can be said to "make management of the earth's carbon cycles through crop cultivation seem like a dismal prospect"—indeed, to make BECCS itself "an exercise in self-delusion".[49] Conscious that a resounding case for the climate benefits of BECCS will require overcoming at least some of these contradictions, many protagonists today therefore place considerable emphasis on the potential benefits of so-called "advanced" (sometimes referred to as "second-generation" or even "third-generation") bioenergy feedstocks. As discussed in Chapter 4, definitions of "advanced" bioenergy vary, but can often refer to processing techniques capable of converting inedible, lignocellulosic plant matter (and not just plant sugars and oils) into combustible forms, as well as to genetic interventions intended to yield increased volumes of sugar or oil, for instance by enhancing the process of photosynthesis itself. Whatever the precise definition, however, the broader vision of advanced bioenergy is frequently one in which so-called idle or marginal agricultural lands will be "improved" through the cultivation of new, highly productive feedstocks, whose tolerance for poor soils and resistance to other environmental stresses will enable them not only to thrive in the absence of chemical fertilisers and irrigation, but also to avoid competing for land with conventional food production or other important uses.[50]

In the UK, the appeal of using advanced bioenergy as a basis for BECCS can be traced not only to the prospect of overcoming some of the innate inefficiencies and environmental contradictions associated with "first-generation" biofuel production, but also to the possibility of becoming less reliant on imported feedstocks, particularly as global demand steadily increases. In concerted efforts to scope out a pathway for the UK to reach net zero GHG emissions by 2050, for example, the faith placed in a new generation of dedicated energy crops is striking. The Committee on Climate Change has recently estimated that drastically expanding the area of domestic land given over to cultivating dedicated energy crops, from "very low levels today" to approximately seven thousand square kilometres, would enable the UK to achieve annual negative emissions of approximately 11 million tonnes of CO_2 per year by the middle of this century.[51] The Royal Society and Royal Academy of Engineering, meanwhile, estimate that annual negative emissions of around 15 million tonnes of CO_2 per year could be achieved by devoting ten thousand square kilometres of land to grow dedicated perennial energy crops, such as *Miscanthus* and short-rotation coppice.[52] When combined with "fully utilising the UK's biogenic waste sources"—such as waste straw from agriculture and residues from forestry—scaling up the domestic cultivation of dedicated energy crops in this way would, the Committee on Climate Change suggests, enable the UK to reduce its current high dependence upon imported biomass feedstocks to below 20 per cent of its total need by 2050.[53]

If such a transformation is to be achieved in practice, of course, significant technical and logistical hurdles will need to be overcome. But there will also be considerable social challenges as well. Wider publics may well be concerned, for example, about the potential environmental impacts and land-use conflicts associated with ramping up domestic energy crop production for BECCS. And there could also be deeper questions raised about the precise role that these crops are ultimately intended to play both within wider UK climate change policy, and as part of a broader process of energy transition away from fossil fuels. That public concerns about BECCS will need to be taken seriously has been underlined starkly by the findings of the UK Citizens' Assembly on Climate Change, undertaken in 2020.[54] Asked to debate the relative merits and drawbacks of diverse possible GHG removal methods that the UK could theoretically use to reach a net zero emissions target by 2050, the assembly's 108 participants expressed less enthusiasm for BECCS than for any of the other five options available (see Table 5.1). Indeed, only 42 per cent of assembly members agreed that the technology

Table 5.1 UK Climate Assembly member preferences for the inclusion of six possible greenhouse gas removal methods in UK net zero climate policy (%).

Option	Strongly agree	Agree	Don't mind / unsure	Disagree	Strongly disagree
Forests and better forest management	81	18	1	0	0
Restoring and managing peatlands and wetlands	50	35	13	2	0
Enhancing the storage of carbon in soil	24	38	23	12	3
Using wood in construction	42	40	15	1	3
Bioenergy with carbon capture and storage	7	35	22	18	18
Direct air capture and storage	11	31	19	21	18

Note: The content on Climate Assembly UK is licensed under the Open Government Licence v3.0 except where otherwise stated.

Source: Author's table, derived from data presented in Climate Assembly UK, *The Path to Net Zero*, 427.

should feature at all in UK plans to reach net zero, and of those, only 1 in 6 classified their agreement as "strong". By contrast, levels of support for using more wood in the construction industry, for restoring and better managing peatlands and wetlands and for planting and better managing forests were all above 80 per cent.

A growing body of empirical research in the social sciences has itself sought to explore public perceptions of bioenergy with CCS, identifying persistent concerns about how and where feedstocks will be produced, who might benefit from the development of new plant cultivars and other forms of intellectual property associated with "advanced" bioenergy supply chains, and how to ensure that future take-up of the technology is incentivised as fairly and cost-effectively as possible.[55] While a shift to a more self-sufficient bioenergy production regime in the UK could allay public concerns about imported feedstocks, such research suggests that the broader social legitimacy of BECCS will nonetheless continue to depend heavily upon citizens' ability to meaningfully shape the terms on which the technology is deployed in practice.[56] On one level, building trust will almost certainly require finding ways to make the geographies and materialities of expanding bioenergy systems more visible and transparent, giving citizens a better idea of exactly where the UK's bioenergy, and indeed where its future negative GHG emissions, will ultimately come from. But it will also require a recognition—on the part of the scientific community, industry actors and policymakers alike—that BECCS is not a ready-made technological solution whose appeal can be defined in purely technical or economic terms, separate to the "messy, situated socio-economic and political contexts" within which it must be deployed in practice.[57] In short, sensitivity to the diverse local contexts within which BECCS must be incentivised, made to function effectively and indeed imbued with wider public legitimacy—not as a standardised technology, but as a series of "coupled socio-technical systems"—will be essential.[58]

From BECCS to BECCUS: turning pollution into the means of production?

While BECCS certainly has the potential to radically transform the materiality and the geography of twenty-first-century global energy systems, its most potent political effects may reside ultimately in the more intangible realm of ideas. Motivated by the prospect of achieving negative GHG emissions, protagonists of BECCS envisage the vital capacities of plants and trees—to photosynthesise and to grow—as a basis for the rapid, large-scale production of organic compounds which can be easily processed into direct replacements for fossil fuels. By contrast, and as discussed earlier in this chapter, the capacities of vegetal lifeforms to keep carbon dioxide locked up in terrestrial ecological settings—central to a view of large-scale tree planting as a "natural climate solution", for example—are cast as inherently inefficient and susceptible to erosion through future climate change itself.[59] CCS technology, when coupled to biomass combustion, promises to replace—and improve upon—these particular vital capacities of plants by storing carbon dioxide safely underground for far longer periods of time than any plant or tree could manage at the earth's surface. From this perspective, BECCS might even be said to render long-term carbon storage in forests, grasslands and other terrestrial ecosystems increasingly redundant as a climate change mitigation tool. Yet this strategic disaggregation of plants' vital capacities—to photosynthesise and to grow on the one hand, and to store carbon dioxide on the other—does more than merely unsettle conventional understandings of the role that vegetal lifeforms should play in regulating the global climate system, as was discussed at the end of Chapter 3. More fundamentally, it also enables plants to be enrolled as a new ecological foundation upon which to strengthen broader relationships between energy consumption, growth and indeed economic value production as a whole.

At one level, these shifts are presaged by the basic logic of BECCS, which dictates that an increase in the burning of hydrocarbon fuels will—so long as CCS technology proves effective—serve not to exacerbate, but rather moderate, the effects of anthropogenic climate change. If it is to work at all, however, CCS must first be rendered affordable as a technology. Unfortunately for enthusiasts of BECCS, efforts to deploy CCS on a large scale have to date been routinely hampered by the costs associated with developing viable technologies and infrastructures.[60] Perhaps unsurprisingly, therefore, hopes are increasingly pinned not strictly on the promises of combining bioenergy with CCS, but rather with carbon capture, *utilisation* and storage (or BECCUS for short). The idea of this move, instead of relying on the future profitability of the long-term geological storage of carbon, is to instead capitalise on value chains for carbon dioxide which have already been established in other economic sectors. While it may seem like a subtle shift, the ramifications of moving from BECCS to BECCUS could be huge. Indeed, protagonists envisage diverse potential applications for captured carbon dioxide, ranging from using the gas to help accelerate tomato growth and boost yields in horticultural glasshouses, through to converting it into polymers and polyols that could then be used as the basis for oil-free alternatives to plastic in the manufacture of diverse commodities, including clothing, furniture and even cars.[61] To be sure, efforts to put such ideas into practice are presently embryonic—the recent manufacture by one energy firm of footwear from captured carbon, for example, generated just five pairs of what was glibly described as the "shoe without a footprint".[62] Yet the ambitions attached to such ideas could hardly be more grandiose. Pointing to research that has converted captured carbon dioxide into an alternative construction material to replace conventional concrete, for instance, Drax have recently suggested that "if the shoes people walk around on can be made from captured carbon, so can the cities they walk within".[63] No less far-reaching, meanwhile, are aspirations attached to the same company's partnership with start-up firm Deep Branch Biotechnology, whose proposed use of captured carbon to produce proteins for use in fish and animal feed aims not just to reduce emissions from the agricultural sector, but also to "help meet the anticipated increase in global demand for meat products".[64]

These diverse efforts at "turning pollution into possibilities" need not necessarily be dependent upon the combustion of biomass fuels for harnessing captured carbon, of course.[65] Yet, while vast amounts of carbon dioxide could theoretically be furnished by the prolific burning of coal, in the UK and EU contexts at least, it seems far more probable that large-scale carbon utilisation and indeed storage will, if they materialise at all, be appended principally to bioenergy generating facilities. Not only are regulatory (and indeed broader societal) moves intended to devalue and phase out coal-based electricity generation already highly developed in these settings, but so too are formal policy commitments to achieving substantial quantities of negative emissions even before the year 2050, when net zero GHG emissions targets are intended to be reached.[66] Regardless of how efficient technologies designed to capture carbon dioxide from power station flue gases may yet become, moreover, coal burning will never be able to serve as a basis for achieving negative GHG emissions. In light of these various biophysical and structural political–economic constraints, the goal of "turning pollution into possibilities" therefore comes to hinge almost exclusively upon the work that plants do to make immediate use of solar energy available at the earth's surface, rather than upon subterranean stocks of solar energy contained in fossil fuels and accumulated in so-called "deep time".[67] The ground upon which these visions are built, consequently, is the establishment of vegetal economies wherein the work of plant growth, and more specifically the "vegetal labour" of photosynthesis and carbon sequestration,[68] serves to bring into being new, ostensibly wasteless metabolic interdependencies between large-scale energy generation and consumption on the one hand, and a growing range of wider forms of commodity production on the other.[69] Bioenergy, in other words, becomes more than just an "input" (whether as heat or as electricity) for various capitalist production processes predicated upon "changing the form of matter", as Marx would have put it.[70] In these visions, it also becomes a means of producing a new "material substratum"—biogenic carbon—whose own form can then be changed within other commodity production processes as well.[71]

Despite being central to BECCUS, however, the precise nature of the vegetal metabolic activities which will actually undergird large-scale biomass cultivation in practice are typically glossed over by protagonists. The photosynthetic work performed by plants and trees rarely becomes anything more than an abstract entity residing somewhere in the background of these visions—a theoretically inexhaustible, natural reserve of "vital force" whose end product, understood simply as biomass, is apprehended strictly at the aggregate level as a continuously self-replenishing reservoir from which to derive renewable fuel.[72] Given the many issues associated with large-scale biomass production that were outlined in the previous subsection, this reticence may not be especially surprising. Yet the ability that plants afford large-scale energy providers to "go negative" in GHG emissions terms, whether through capturing and storing carbon dioxide or indeed putting that gas to use as a basis for other forms of commodity production, is nonetheless at the core of ongoing and unabashed efforts to recast energy consumption itself as a basis not just for economic growth, but also for climate repair. Accordingly, Drax—along with a recently formed consortium of wider industrial actors—advocate decisive government investment in BECCUS as an essential means not merely of drastically reducing the GHG emissions associated with UK electricity generation, but also of helping other energy-intensive industries based in the nearby Humber Estuary region (such as refining, petrochemicals and steel manufacturing) to "continue to operate and thrive even against a backdrop of tighter emissions targets".[73] Indeed, for this so-called "Zero Carbon Humber" consortium, investment in the necessary infrastructures to achieve BECCUS at Drax will exert a galvanising effect upon the regional economy as a whole, not least by rendering investment in CCS technology in other industries—including most notably hydrogen fuel production—more economically viable.[74] The prospect of creating tens of thousands of jobs, not to mention generating significant wider investment, is here attached inextricably to the switch from an energy system predicated substantially upon coal burning to one predicated increasingly upon large-scale biomass combustion.[75] Indeed, by creating the "world's first zero carbon industrial cluster", the consortium even suggests that the region could serve as a blueprint for wider economic revitalisation, "taking the UK from a green recovery to a world-leading industrial powerhouse".[76]

Much as such visions depict BECCUS as a vital component of wider decarbonisation strategies in the UK, however, any "clean growth" that it is to achieve will inevitably act to perpetuate, rather than disrupt, the now centuries-long practice of generating electricity from the large-scale burning of hydrocarbon fuels.[77] To cast BECCS (or BECCUS) as a decarbonisation tool for the UK economy in the same vein as solar arrays or wind farms, therefore, serves only to obfuscate the precise grounds upon which the technology has garnered such enthusiastic support from incumbent energy actors to date. The real value of BECCS, I argue, lies not strictly in any inherent capacity to decarbonise the UK's energy system, so much as in its potential to meticulously and precisely "recarbonise" that system—in other words, to restructure it around novel (and ostensibly wasteless) metabolic pathways very different to those involved in coal burning.[78] However seamlessly visions of BECCS might gloss over their vegetal foundations, moreover, it is plant work, or more specifically the "elemental rearrangement" that plants achieve through photosynthesis,[79] which serves as the inescapable basis of this "recarbonisation" process. Indeed, following Holly Jean Buck, BECCS might even be described as an attempt to reorient agricultural and forestry systems "toward a kind of carbon alchemy: a sophisticated transmutation of life and land for the ends of sequestering carbon".[80] Yet the long-term sequestration of carbon dioxide underground and, moreover, the use of biogenic carbon as a means of production in its own right are not the only end goals of these visions. Rather, by deliberately using plants to channel carbon dioxide out of the atmosphere and then converting this carbon into new forms elsewhere (whether as diverse "sustainable" commodities or vast underground reservoirs of "captured" emissions), protagonists of BECCUS aim, above all else, to press the vegetal world into service as a theoretically inexhaustible basis for the renewed pursuit of perpetual economic growth. In other words, at its most fundamental level, BECCUS is ultimately a project concerned—I argue—not with decarbonisation at all, but rather with shoring up and consolidating societal understandings of intensive energy generation, large-scale industrial activity, and resource consumption as valuable and indeed virtuous acts in their own right.

Fanning the flames? Exposing the fossil logics of BECCS

Enthusiasts of BECCS proclaim the coupling together of biomass combustion with CCS technology as a progressive, "game changing" move in the context of wider efforts to tackle anthropogenic climate change. Far from acting simply to reduce the GHG emissions associated with existing forms of large-scale energy generation (as when CCS is invoked as a means of delivering so-called "clean coal", for example[81]), BECCS promises instead actively to remove carbon dioxide from the atmosphere on an effectively permanent basis. By presenting an overview of the history of the idea of BECCS, this chapter has sought to scrutinise more carefully the precise influence of the technology on the contemporary politics of climate change. At one level, the chapter views BECCS as an archetypal "socioecological fix" for an impending crisis of capital in the energy sector, one which functions by staving off the imminent devaluation of incumbent forms of accumulation based on fossil fuel burning, rather than by addressing the biophysical root causes of climate warming head on. In other words, through its promise to realise vast quantities of negative GHG emissions in the coming decades, BECCS is already serving to postpone the point at which ongoing practices of large-scale fossil fuel burning will become economically unviable. From this perspective, and as geographer Wim Carton puts it, BECCS can be viewed as "the mobilisation of a specific vision of the future as a way to legitimise and reproduce the present".[82] Yet the chapter has also insisted that BECCS "reproduces" the present not just by delaying the devaluation of incumbent fossil fuel-based energy generating infrastructures. Instead, at a more fundamental level, by staking humanity's hopes of achieving significant negative GHG emissions this century upon a transition from fossil fuels to biomass fuels, BECCS also acts to perpetuate broader systems of energy generation predicated upon the large-scale, centralised combustion of hydrocarbon fuels. Bold commitments to BECCS, in short, are inherently commitments also to sustaining pyrogenic energy generation practices as a dominant motor of economic growth in many developed economies of the global North, regardless of whether fossil fuel reserves themselves are ever eventually abandoned or not.

Whether efforts are made to combine biomass combustion with carbon capture and long-term storage (in order to effectively substitute naturally occurring terrestrial carbon sinks with deliberately engineered, subterranean ones), or to combine biomass combustion with carbon utilisation (thereby turning pollution into a means of production), this chapter therefore advocates a rejection of the idea that BECCS constitutes a tool for decarbonising incumbent energy systems. Instead, given its fundamental reliance upon the photosynthetic work of vegetal lifeforms, I have argued that BECCS is a project concerned ultimately with the deliberate, controlled "recarbonisation" of resource-intensive economies, especially those located in the global North. Though their work is only ever invoked at an abstract, aggregate level, plants are here indisputably viewed as the fundamental ecological foundation for future forms of "clean growth"—such as those envisaged by the Zero Carbon Humber consortium—to be achieved through BECCS. The paradox here is that BECCS promises to decouple large-scale energy generation from the twin problems of resource depletion and climate change not by making greater use of plants and trees as carbon sinks (as is the case where tree planting is portrayed as a natural climate solution), but rather by pressing plants and trees into service as increasingly intensive *producers* of GHG emissions. Finding ways to permanently store those emissions, or indeed to feed them into either existing or new production processes in innovative ways, is then up to humanity. What is being proposed, from this perspective, is nothing less than the establishment, at the planetary scale, of collaborative regimes of intensive carbon metabolism in which the vital capacities of vegetal lifeforms will be increasingly intertwined with, and deliberately improved and complemented by, human scientific, technological and engineering capabilities. Even as it seeks to help realise "the ideal of a zero-waste humanity"—an ideal that is at the core of wider visions of a fully bio-based, circular economy—BECCS will inevitably also entail putting an unprecedented proportion of the living nonhuman world to work.[83] In short, BECCS therefore "invokes the logic of subsumption on a planetary scale", proposing to harness the vegetal labour power of plants to stabilise the climate by furnishing economies with not just abundant volumes of fuel but also carbon dioxide itself.[84]

Of course, the sheer ambition of visions associated with BECCS could hardly contrast more sharply with the modest success of attempts, to date at least, to put such projects into practice "on the ground". This stark contradiction, however, is ultimately overridden by the promise of BECCS to help solve climate change without calling into question the resource-intensive lifestyles which have to date been enjoyed only by (some of) those living in the global North. Indeed, the full promise of BECCS arguably goes further even than this, by raising the prospect that intensive energy consumption could—if undergirded by plant resources rather than fossil ones—serve actively to repair the damage that historical anthropogenic GHG emissions have already done to the global climate system. Viewed in this light, BECCS constitutes a key plank of what Jesse Goldstein has diagnosed as a wider "net zero ideology"—a strategy which defers meaningful transformations to incumbent energy systems, even as it reassures populations in the global North that existing forms of modernity "need not be undone in pursuit of a just and viable future".[85] Any concerns that may be raised about the huge challenges—not to mention the diverse possible negative environmental and social impacts—of enacting large-scale bioenergy production in specific places, come from this perspective to be trivialised as mere teething problems rather than as universal objections to the enticing idea that BECCS is humanity's ticket to a "perpetual present of material and energetic abundance".[86] To expose the full political stakes of BECCS—I argue—requires a focus on the local experiences and situated effects of intensive bioenergy feedstock cultivation to be transcended by a broader critique of the relationships that this technology seeks to establish between vegetal lifeforms, energy, value and human progress. Attention must be paid, in short, to the novel vegetal political economies that BECCS, and BECCUS, are already beginning to bring into being.

On this final point, the chapter therefore concludes by suggesting that dominant visions of BECCS today—whether outlined by political institutions, scientific and expert advisory bodies or powerful energy interests—must in the end be recognised as highly developed incarnations of today's prevailing regulatory logic of modern bioenergy, one which aspires to lay claim to the vegetal kingdom writ large as a living, supposedly natural embodiment of the fundamental economic principles of efficiency, productivity and ultimately growth itself. After all, despite promising to enable negative GHG emissions, BECCS as currently envisaged still fails—in common with all modern bioenergy projects—to meaningfully challenge what critical environmental scholar Larry Lohmann has called the "fossil inheritance" of prevailing societal ideas about the nature and purposes of energy.[87] Instead, by invoking hard-working plants and trees as a theoretically inexhaustible basis for achieving genuinely "clean growth", protagonists of BECCS reinforce a particular, European historical narrative of human relationships to energy and fuel, one whose origins can be traced directly to the advent of large-scale fossil fuel burning and the development of the "laws" of thermodynamics in the nineteenth century.[88] The central premise of this narrative—or, to describe it in Cara Daggett's terms, this "fossil myth"—is an ethos of "energy expansion-as-progress", wherein humanity's success as a species is depicted as fundamentally reliant on its ability to fulfil an apparently unquestioned demand for ever-increasing volumes of energy.[89] Against this backdrop, my argument is that it will be necessary—if current visions of BECCS are to be effectively challenged—to fully expose and confront the presumption, central to today's prevailing regulatory logic of modern bioenergy, that it is simply in human nature to desire ever more energy, and moreover that human progress should necessarily entail the continual expansion of available energy resources, even despite climate change and other forms of environmental degradation.[90] With this in mind, the next and final chapter advocates a return to the literal roots of modern bioenergy logics—plants themselves—whose photosynthetic exertions are presently all-too-often backgrounded within dominant imaginaries of BECCS. More specifically, the book's conclusion asks how a closer attunement to vegetal lifeforms might yet serve, if we allow ourselves to be open to new ways of thinking, to enable genuinely post-fossil ideas about energy, and indeed energy's relationships to human progress, to come into being and flourish.

Notes

1 Drax Group plc, "Carbon dioxide now being captured in first of its kind BECCS pilot."
2 For a discussion of the promises and pitfalls of carbon capture and storage research and policy, see Scott *et al.*, "Last chance for carbon capture and storage"; and Bäckstrand *et al.*, "The politics and policy of carbon capture and storage."
3 For a fuller discussion of expanding bioenergy resource-making practices predicated upon working forests of the US South, see Chapter 3.
4 See Office for National Statistics, "A burning issue." As discussed in earlier chapters, GHG emissions generated by burning biomass fuels in the UK (and elsewhere) are officially counted as zero, on the assumption that they will subsequently be offset by carbon sequestration achieved within the land-use sectors of countries where biomass feedstocks are cultivated.
5 Drax Group plc, "Carbon dioxide now being captured in first of its kind BECCS pilot."
6 Committee on Climate Change, *Net Zero*; and Royal Society and Royal Academy of Engineering, *Greenhouse Gas Removal.*
7 Smith *et al.*, *The State of Carbon Dioxide Removal*, 61.
8 Royal Society and Royal Academy of Engineering, *Greenhouse Gas Removal*, 101.
9 Intergovernmental Panel on Climate Change, *Climate Change 2022*, 25.
10 Intergovernmental Panel on Climate Change, *Global Warming of 1.5°C.*
11 Hickman, "How BECCS became climate change's 'saviour' technology"; and Chatham House, *Net Zero and Beyond*, 4.
12 Hickman, "How BECCS became climate change's 'saviour' technology."
13 Obersteiner *et al.*, *Biomass Energy, Carbon Removal and Permanent Sequestration*, iii.
14 Obersteiner *et al.*, *Biomass Energy, Carbon Removal and Permanent Sequestration*, 31.
15 Obersteiner *et al.*, *Biomass Energy, Carbon Removal and Permanent Sequestration*, 30.
16 Obersteiner *et al.*, *Biomass Energy, Carbon Removal and Permanent Sequestration*, 6.
17 For a fuller discussion of the genealogy of this idea of an environmental "sink", see Gabrys, "Sink: The dirt of systems."
18 Obersteiner *et al.*, *Biomass Energy, Carbon Removal and Permanent Sequestration*, 2 (emphasis added).
19 Intergovernmental Panel on Climate Change, *Global Warming of 1.5°C.* For critical assessments of these developments, see Beck and Mahony, "The politics of anticipation"; and Cointe and Guillemot, "A history of the 1.5°C target."
20 For a fuller history of this target, see Randalls, "History of the 2°C climate target."

21 Intergovernmental Panel on Climate Change, *Climate Change 2014 Synthesis Report*. See also Smith and Porter, "Bioenergy in the IPCC assessments."
22 Chatham House, *Net Zero and Beyond*, 5 (emphasis added); see also Malm and Carton, *Overshoot*.
23 Wim Carton, "'Fixing' climate change by mortgaging the future," 750; and Goldstein, "Net zero and settler futurity," 33.
24 The *Synthesis Report* of the IPCC's 2014 Fifth Assessment Report admits that "many models could not limit *likely* warming to below 2°C if bioenergy, CCS and their combination (BECCS) are limited (*high confidence*)" (Intergovernmental Panel on Climate Change, *Climate Change 2014 Synthesis Report*, 24, emphasis in original).
25 Hickman, "How BECCS became climate change's 'saviour' technology."
26 Anderson and Peters, "The trouble with negative emissions," 183.
27 See for example Bellamy *et al.*, "Incentivising bioenergy with carbon capture and storage (BECCS) responsibly."
28 Carton, "'Fixing' climate change by mortgaging the future," 762.
29 Carton, "'Fixing' climate change by mortgaging the future," 762; on the politics of discount rates in IAMs, see van Vuuren *et al.*, "Open discussion of negative emissions is urgently needed."
30 Doganova, *Discounting the Future*.
31 Beck and Mahony, "The politics of anticipation," 1. The phrase "political economy of delay" was coined in Carton, "'Fixing' climate change by mortgaging the future." See also Dooley *et al.*, "Co-producing climate policy and negative emissions."
32 McCarthy, "A socioecological fix to capitalist crisis and climate change?" 2495.
33 Carton, "'Fixing' climate change by mortgaging the future."
34 C-Capture, "£5M boost to scale up ground-breaking carbon capture pilot at Drax."
35 Energy Technologies Institute, *Bioenergy*, 3.
36 Geden, "The Paris Agreement and the inherent inconsistency of climate policymaking."
37 Rayner, "What might Evans-Pritchard have made of two degrees?"; see also Kruger *et al.*, "Abandon hype in climate models."
38 Huber and McCarthy, "Beyond the subterranean energy regime?"; see also Huber, "Energized antagonisms: Thinking beyond 'energy culture'"; and Malm, *Fossil Capital*.
39 See Dunlap, *Renewing Destruction*; and Boyer, *Energopolitics*.
40 Smith *et al.*, "Biophysical and economic limits to negative CO_2 emissions," 46.
41 Dooley *et al.*, "Over-reliance on land for carbon dioxide removal in net-zero climate pledges," 4.
42 Royal Society and Royal Academy of Engineering, *Greenhouse Gas Removal*, 109.
43 Buck, *After Geoengineering*, 62; see also Geden *et al.*, "Integrating carbon dioxide removal into EU climate policy"; and Hickel, "The anti-colonial politics of degrowth."

44 See Bellamy and Palmer, "Geoengineering and geographers."
45 Ricardo Energy and Environment, *Analysing the Potential of Bioenergy with Carbon Capture in the UK to 2050*, 9.
46 See for instance Borras *et al.*, "The politics of biofuels, land and agrarian change"; Buck, *After Geoengineering*; and Nally, "Governing precarious lives."
47 For a fuller discussion of the social and environmental issues associated with biofuel-driven land-use change, see Chapter 2.
48 Fajardy and MacDowell, "Can BECCS deliver sustainable and resource efficient negative emissions?"
49 Buck, *After Geoengineering*, 64.
50 Designations of "marginal" lands or "wastelands" as requiring "improvement" are of course contentious, and have historically been bound up—as Jesse Goldstein has pointed out—with processes of primitive accumulation and the production of "enclosed" forms of nature. See Jesse Goldstein, "*Terra economica.*"
51 Committee on Climate Change, *Net Zero*, 147; see also Committee on Climate Change, *Land Use*, 8–9.
52 Royal Society and Royal Academy of Engineering, *Greenhouse Gas Removal*, 101.
53 Committee on Climate Change, *Net Zero*, 151.
54 Climate Assembly UK, *The Path to Net Zero.*
55 Bellamy *et al.*, "Incentivising bioenergy with carbon capture and storage (BECCS) responsibly"; and Waller *et al.*, "Contested framings of greenhouse gas removal and its feasibility."
56 In this respect, work on public perceptions of BECCS echoes many of the foundational tenets of longer-standing bodies of work—in geography and science and technology studies, especially—examining the root causes of public controversies around emerging technologies, and broader interactions of knowledge, expertise and trust within environmental policymaking. See for example Owens, "'Engaging the public'"; *Owens, Knowledge, Policy, and Expertise*; and Eden, *Environmental Publics.*
57 Bellamy and Palmer, "Geoengineering and geographers," 528.
58 Bellamy *et al.*, "Perceptions of bioenergy with carbon capture and storage in different policy scenarios," 2.
59 See for instance Forzieri *et al.*, "Emerging signals of declining forest resilience under climate change."
60 See for example Scott *et al.*, "Last chance for carbon capture and storage."
61 Drax Group plc, "What can be made from captured carbon?"
62 Varinsky, "An energy company just debuted a shoe made from power plant emissions."
63 Drax Group plc, "What can be made from captured carbon?" On the production of sustainable concrete, see Foulsham, "UCLA researchers turn carbon dioxide into sustainable concrete."
64 Drax Group plc, "New carbon capture technology could help industry and agricultural sector decarbonise."

65 Drax Group plc, "What can be made from captured carbon?"
66 For a discussion of the varying efficacy of regulatory attempts to promote coal phase-out in Europe, see Rentier *et al.*, "Varieties of coal-power phase-out across Europe." On proactive efforts to "devalue" fossil fuel-based energy generation technologies and infrastructures, see Knuth, "Green devaluation: Disruption, divestment, and decommodification for a green economy."
67 See for example Jonsson, "Abundance and scarcity in geological time."
68 For a fuller discussion of vegetal productivity as a form of labour, see Chapter 1 and Palmer, "Putting forests to work?"
69 This vision even extends, in some scenarios, to the production from captured carbon dioxide, using emergent fuel cell technologies, of new types of fuel itself, thereby enabling bioenergy to become an energy resource which can effectively be consumed twice. See Drax Group plc, "Drax secures £500,000 for innovative fuel cell carbon capture study."
70 Marx, *Capital: A Critique of Political Economy*, 1:133. Marx argues that "the use values ... of commodities, are combinations of two elements—matter and labour. If we take away the useful labour expended on them, a material substratum is always left, which is furnished by Nature without the help of man. The latter can work only as Nature does, that is by changing the form of matter."
71 Marx, *Capital: A Critique of Political Economy*, 1:133. The future value of biogenic carbon as a "means of production" in other economic sectors will depend critically, of course, on the success of efforts to scale up BECCUS itself. Without a drastic increase in supply, biogenic carbon is likely to remain economically uncompetitive against alternative resource inputs for the production of synthetic fuels, such as green hydrogen.
72 For Miriam Boyer and colleagues, the appeal of this abstract notion of biomass derives from its capacity to render "the complexity and constant transformations associated with living nature ... tame and controllable" (Boyer *et al.*, "The making of sustainability," 683).
73 Zero Carbon Humber, *Capture for Growth*.
74 See Zero Carbon Humber, *Capture for Growth*.
75 See Vivid Economics, *Capturing Carbon at Drax*.
76 Zero Carbon Humber, *Capture for Growth*; and Drax Group plc, "Jobs, skills, zero emissions."
77 Vivid Economics, *Capturing Carbon at Drax*.
78 Palmer and Carton, "Carbon removal as carbon revival?" 3.
79 Myers, "Photosynthetic mattering," 123.
80 Buck, *After Geoengineering*, 57.
81 Brown and Spiegel, "Coal, climate justice, and the cultural politics of energy transition."
82 Carton, "'Fixing' climate change by mortgaging the future," 764.
83 Zwier *et al.*, "The ideal of a zero-waste humanity."
84 Carton *et al.*, "Revisiting the 'subsumption of nature'," 792.
85 Goldstein, "Net zero and settler futurity," 32.

86 Goldstein, "Net zero and settler futurity," 45.
87 Lohmann, "Bioenergy, thermodynamics and inequalities," 9.
88 Cara New Daggett, *The Birth of Energy*.
89 Cara New Daggett, "Energy and domination," 647.
90 For a discussion of the inherent philosophy of growth and expansion buried within visions of supposedly circular, bio-based economies, see Zwier *et al.*, "The ideal of a zero-waste humanity."

Conclusion: rethinking plant power—for vegetal energy cultures

> The techno-capitalist framing of plants de-eventalizes them.
>
> —Michael Marder, *The Sense of Seeds, or Seminal Events*, 2015

> You will never be master of the tree. All you can ever wish for is to become a tree in your turn.
>
> —Georges Perec, *A Man Asleep*, 1967

Modern bioenergy is, at bottom, nothing less than planty pyromania. The real purpose of burning plants on an industrial scale is not—this book has argued—to gradually prune global fossil fuel consumption, thereby addressing the twin crises of climate change and energy security. Rather, modern bioenergy must be understood, I have contended, as a broader regulatory logic concerned ultimately with shoring up energy cultures which have been enjoyed predominantly by affluent nations of the global North during the fossil fuel era. These are energy cultures inextricably bound up, unfortunately, not just with industrialisation but also with violent, colonising regimes of land dispossessions, resource extractivism and economic oppression in the less affluent global South. Even if modern bioenergy were to be successful in decoupling large-scale hydrocarbon combustion from global climate warming, moreover, on a more fundamental level, it would nonetheless continue to perpetuate a "fossil myth" linking ever more efficient and productive energy use to human progress writ large.[1] Against these prevailing ideas, a core assertion of this book has also been that today's regulatory logic of modern bioenergy—despite some stunning successes—remains inherently fragile. By enrolling vast swathes of forest and agricultural land into the production of fuel, modern bioenergy quite literally unearths processes of industrial hydrocarbon provisioning that were hitherto confined to the subterranean realm. In substituting the vertical for the horizontal, the dead for the living, modern bioenergy's protagonists bank on plants themselves to help naturalise the economics, politics and cultures of resource use which have shaped the era of fossil energy. There is nothing natural, however, about enrolling the metabolic "work" of plants—their capacities to photosynthesise, grow and sequester carbon dioxide—into large-scale energy resource-making. Indeed, by insisting upon a view of "the work that plants do" in these contexts as "vegetal labour", I have sought to foreground the contingency and contestability of the end uses towards which plant metabolism is currently being put in the ongoing modern "bioenergy boom".[2] Building further on these arguments, the central aim of this final chapter is to outline how modern bioenergy might yet serve, if we should choose to more fully open ourselves up to the influence of plants, to provoke a radical rethink of prevailing energy–society relationships—relationships which are currently presumed to link intensive, continuously expanding energy consumption to perpetual economic growth and, by extension, to human achievement writ large. Before embarking on this task, however, it is first necessary to enumerate more systematically a series of structural tensions already latent in today's regulatory logic of modern bioenergy—tensions which highlight the inherent susceptibility of this logic to being reconfigured and reoriented towards different end goals.

Modern bioenergy: juggernaut or house of cards?

Throughout this book, I have insisted that modern bioenergy is not simply a discrete object of climate and energy governance regimes, functioning as one among many renewable resource inputs to be fed into existing social and economic systems. Instead, I have conceptualised modern bioenergy as a set of guiding principles and ideas for enrolling plants into the restructuring and renewal of relationships which imbue intensive energy consumption with both economic and cultural legitimacy in the first place. The translation of these guiding principles and ideas into materially existing versions of modern bioenergy, of course, has not been systematic or uniform across all parts of the globe. Modern bioenergy, in other words, is not a coherent political project, but rather a more disparate governance "apparatus" comprising, in Foucault's terms, an ensemble of "discourses, institutions, architectural forms, regulatory decisions, laws, administrative measures, scientific statements, [and] philosophical, moral and philanthropic propositions".[3] This apparatus, moreover, is an inherently fragile construct; a "system of relations" which must be made and remade over time through the utterances and actions not of any single, globally hegemonic authority or actor, but rather of countless organisations and individuals invested in the day-to-day work—cognitive, discursive, scientific and political—of mobilising vegetal labour to help produce combustible hydrocarbon fuels.

Even if it is theoretically fluid, always in-the-making and inherently open to change, however, the dominant strategic function of this governance apparatus as it is currently configured has been to enrol ever more kinds and ever greater volumes of plant life to take on the work—or labour—previously performed by fossil fuels, not simply as energy carriers but also as sources of productive political power. Indeed, and as I have sought to show in the preceding four chapters, today's modern bioenergy apparatus has already served not merely to replace an appreciable volume of fossil fuel use in sectors like road transport and electricity generation, but also to structurally transform energy–society relationships in the global North on three distinct levels. These transformations can be witnessed, I have argued, in the emergence of new geographical orderings of hydrocarbon provisioning at the international scale, in the creation of new kinds of vegetal natures intended to mimic fossil fuels as closely as possible, and ultimately in the enrolment of plant metabolism—as vegetal labour power—into the creation of new forms of economic value arising as much from global climate stabilisation as from energy generation itself. The successes that modern bioenergy's protagonists have achieved in the relatively short period since the initial new biofuel rush of the early 2000s should not, therefore, be underestimated. Injecting well over 300 million litres of oil-equivalent bioethanol and biodiesel into global circuits of liquid hydrocarbons on a daily basis is no mean feat.[4] Similarly, the recent and still-growing replacement—in the UK, northwest Europe, South Korea, Japan and elsewhere—of tens of millions of tonnes of coal each year by wood pellets derived principally from forests in North America represents an unprecedented achievement in the history of forest-based energy production.[5] The more than eight million tonnes of wood pellets which were burned just at the UK's Drax power station in 2021, for example, exceed the total volume of wood burned throughout England around the middle of the eighteenth century by a factor of four.[6] At the same time, however, it is also true that many of the claims and imaginaries advanced by modern bioenergy's enthusiasts remain largely unfulfilled and promissory. The ambitious visions outlined in Chapters 4 and 5, for example, of a net zero American aviation sector undergirded by crop-based SAFs, and of large-scale BECCUS as a means of meeting net zero climate targets both in the UK and at a global scale, remain open to the charge of constituting mere "incantatory" or "magical thinking", despite having received significant volumes of government financial support.[7] This ongoing juxtaposition—of impressive progress in some areas contrasted with, as yet at least, starkly unfulfilled promise elsewhere—hints at a number of more fundamental tensions which, as I set out below, imbue today's prevailing regulatory logic of modern bioenergy with an inherent fragility and openness to critique.

The first of these tensions resides in the tendency of many of modern bioenergy's protagonists, as discussed in Chapter 1, to conflate vitalism with productivism. When pitched against fossil fuels, plants living and growing at the earth's surface today appear to transcend two major problems—those fuels' finiteness and their inextricable links to global climate warming. Plants do, of course, constitute a renewable if not theoretically inexhaustible resource. As living organisms, moreover, it is also true that plants can conjure valuable organic compounds out of "immediate circulations of solar energy" arriving at the earth's surface in real time, thereby offering a pathway for the decoupling of hydrocarbon burning from the release of prehistoric carbon locked up for hundreds of millions of years in the so-called "subterranean forests" of coal, oil and gas.[8] Nonetheless, the "liveliness and mutability" of plants alone are insufficient to explain the widespread scientific, commercial and political appeal of modern bioenergy.[9] Instead, it is the capacity of vegetal lifeforms to physically embody productivity and growth, and moreover to offer up the possibility of further optimising and enhancing that growth (in short, to facilitate the "real subsumption of nature"[10]) that properly underpins depictions of modern bioenergy as a desirable project for humanity today. This latter, explicitly productivist view is epitomised by depictions of the live crowns of loblolly pines as photosynthetic "engines" powering the production of bigger and better trees in commercially managed forests of the US South, just as it is by an insistence, in research communities seeking to produce advanced bioenergy crops for SAF, that plants should be viewed "as factories" geared towards emulating the multifunctional economic capacities of fossil oil itself. These views complicate more commonplace accounts of modern bioenergy's benefits—for energy security and climate mitigation alike—as dependent simply upon the status of plants as vital organisms. In practice, I argue that modern bioenergy's apparent virtuousness derives at least as much, if not more, from the ease with which plants can be upheld as paragons of productivity and growth—from its association, in Andreas Malm and Wim Carton's terms, with fields of "sugarcane or willows or poplars modified to grow so fast the eye can almost see it".[11] In short, modern bioenergy relies upon the vegetal kingdom to materially prefigure intensive hydrocarbon burning as a desirable—indeed, necessary—preoccupation for society, even as moves to abandon fossil fuels themselves ostensibly gather pace.

A second, closely related tension is found in accounts of modern bioenergy as the basis for powering an economy—or, more specifically, a "bioeconomy"—which is both "circular" on the one hand, and yet also capable of sustaining perpetual productivity increases and expansion on the other. As Mario Giampietro has recently shown, the terms "circular economy" and "bioeconomy" are frequently entwined in global North policy discourse, implying that a wholesale shift from fossil resources to biological ones would enable society to "close the loop" between resource consumption and waste generation.[12] For Jochem Zwier and colleagues, however, the ultimate incarnation of circularity—a "zero-waste" economy—would be structurally at odds with any need to increase the volume of solar energy harnessed by plants and converted into biomass over time.[13] Widespread commitments to large-scale modern bioenergy expansion belie this objective, of course. In reality, however, policy commitments to a "circular bioeconomy" do not seek principally to eradicate waste; rather, the central goal is to use specific circular production processes to reinvigorate growth.[14] With respect to the manifestations of modern bioenergy examined in this book, this second tension is perhaps nowhere better explicated than in utopian visions of BECCUS as a means of achieving large-scale negative GHG emissions, discussed in Chapter 5. BECCUS appears at core to be a project concerned with achieving a form of zero- or very low-waste circularity; carbon dioxide emissions generated when biomass fuels are burned will, at worst, be permanently stored in geological rock formations, and at best be harnessed as the "means of production" for other industrial processes, ranging from sustainable concrete production to the manufacturing of bioplastics or even synthetic fuels. The deeper appeal of this idea, however, is precisely that it goes beyond rendering bioenergy "neutral" or harmless from a climate change perspective. Much more radically, BECCUS turns the conventional association between hydrocarbon burning and climate warming on its head, effectively creating an imperative to burn more biomass not just to reduce the rate at which warming progresses, but to actively repair the global climate system. By this logic, the faster and more quickly bioenergy is consumed, the more carbon dioxide will be extracted from the atmosphere, and the more quickly the climate crisis will be solved.[15] Despite their reliance on the rhetoric of circularity, therefore, utopian visions of BECCUS are ultimately committed to the same forms of perpetual energy expansion and economic growth which were normalised by the era of fossil capitalism, an era which modern bioenergy is meant to be leaving behind.

To some extent, these first two areas of tension—vitalism against productivism, and circularity against expansion—might be viewed as rooted in binary distinctions between reproduction and production which become increasingly "tenuous" in the context of vegetal growth.[16] By contrast, the next pair of tensions pertain not to the fundamental modalities of modern bioenergy, but rather to the geographical and temporal forms this regulatory logic has adopted, and indeed continues to adopt, in practice. Geographically, the fundamental tension resides in a juxtaposition between dreams of using "home-grown" biofuels to enhance, even if only modestly, the energy security of affluent economies of the global North, and the steady expansion of lands used for intensive biofuel feedstock cultivation in the global South. This is a tension, in short, between patriotic commitments to harnessing national scientific prowess and agricultural expertise to achieve self-sufficiency in biofuels on the one hand, and—on the other—a meek surrender to the economic and political expediencies of importing vast quantities of combustible biomass feedstocks from "sacrifice zones" found in disparate locations across Latin America, sub-Saharan Africa and South East Asia.[17] In the US especially, policy commitments to ramping up domestic biofuel production have been remarkably persistent even in the face of a "global biofuel scape" characterised by the frenzied international trading of vast volumes of biomass feedstocks.[18] George W. Bush's vision of regular "crop reports" being delivered to future US presidents may have seemed quaint even in 2005, but this has not stopped the DOE from producing numerous modelling studies aiming to showcase the feasibility of producing north of one billion tonnes of biomass on American soil each year before the middle of this century.[19] No effort to establish a territorially delimited national (or even regional) bioenergy production system, however, has ever been genuinely successful. Instead, the astonishing growth in European and American consumption of liquid transport biofuels and solid wood pellets, as charted in Chapters 2 and 3, has ultimately been achieved in ways that sustain an almost perfect invisibility of those forms of modern bioenergy at the point of their use, even as their production has been brutally consequential for diverse societies and ecosystems located conveniently out of sight. Frequently, the land acquisitions necessary to enable large-scale biofuel feedstock cultivation in the global South have been justified through familiar colonial narratives of "improvement" and "wasteland" rehabilitation—narratives which are deployed disingenuously to legitimise the condemnation of these lands, in reality, to a state of "permanent exhaustion and debilitation", all in the name of preserving high-consumption lifestyles in the global North.[20]

The fourth and final tension which defines the prevailing regulatory logic of modern bioenergy today is, then, its ambiguous, liminal position with respect to the inertia of incumbent fossil fuel-based energy infrastructures and calls for a radical transformation, or "transition", to a new kind of global energy system. Modern bioenergy may well be a largely "overlooked giant" of the renewable energy landscape, but perhaps this is for the straightforward reason that it is not meant to be noticed at all.[21] The increased blending of biofuels into conventional petrol and diesel since the turn of the century, as discussed in Chapter 2, may well have acted to displace a reasonable volume of fossil fuel use, but it has also reinforced and sustained road transport infrastructures yoked inextricably to the internal combustion engine and liquid hydrocarbons. In this sense, the new biofuel rush of the early 2000s can hardly claim to have unsettled the fundamental links between driving and the dead labour of prehistoric plants ossified into oil. At best, modern bioenergy might instead be described as having brought into being "stacked energyscapes" in which fossil fuels and biofuels "intersect and complement one another", rather than as having initiated any genuine energy "transition".[22] Put differently, this is a project reinforcing "the entanglement and symbiotic expansion of all energies", rather than the straightforward substitution of dirty fossil fuels for supposedly clean, renewable ones.[23] Fast-forward to the mid-2020s, moreover, and these contradictions have only intensified. Grandiose visions of a climate system stabilised through the future, industrial-scale enrolment of the vegetal labour of crop plants and fast-growing trees into BECCUS systems, despite appearing to acknowledge the pressing need to end fossil fuel use, can be seen in practice as deferring meaningful transformations to incumbent energy systems—all on the presumption that future technologies will be able to reduce atmospheric concentrations of carbon dioxide far more quickly than those available today.[24] In its more up-to-date guises, modern bioenergy has thus become a key accomplice in the wider shift, astutely discerned by Wim Carton and others, from discourses of delay to discourses of deferral in global climate politics.[25] The ultimate point of these discourses, expressed elegantly by Jesse Goldstein, is to reassure societies already benefitting from the trappings of historical capitalist extraction and colonial exploitation that high-consumption lifestyles "need not be undone in the pursuit of a viable and just future that extends abundance at all".[26] By promising at once to provide abundant energy and to permanently capture carbon dioxide (or else put it to productive use in other industries), BECCUS entices us not simply to believe that excess GHG emissions will be easy to remove from the atmosphere again in the future. More fundamentally, it asks us to redouble our commitment to the industrial-scale burning of hydrocarbons—now underwritten by the work of plants rather than by fossils—as a means of eventually correcting the "overshoot" which those commitments have historically produced in the first place. Whether such a future will ever materialise in practice, of course, is a moot point.

Taken together, these four tensions—I argue—render the prevailing regulatory logic of modern bioenergy, despite the grandiosity of its visions and the undeniable scale of its successes to date, inherently fragile and vulnerable to critique. On a fundamental level, modern bioenergy indeed epitomises Scott Prudham's description of all capitalist industries which rely on "nonproduced" biophysical inputs as "curiously dependent and vulnerable, often without our acknowledgement".[27] Modern bioenergy specifically, however, is curiously dependent on "nonproduced" inputs—in its own case, dedicated energy crops and fast-growing trees—not just as a basis for tangible commodity production and economic value creation, but also as metaphorical vehicles for reinforcing prevailing energy cultures—cultures depicting human progress itself as inextricably linked to the ever more efficient and productive consumption of larger and larger amounts of energy. Plants are the elixir, from this standpoint, which will finally enable society to overcome the biophysical limits to growth found in resource depletion and ecological degradation. Yet this is a precarious situation, I argue, because there is in the end nothing natural at all about enrolling ever-expanding and ever more diverse forms of vegetal labour into the project of large-scale hydrocarbon provisioning, or indeed climate stabilisation for that matter. For modern bioenergy's protagonists, plants may be easily depicted as paragons of productivism and perpetual growth, but these depictions at best oversimplify, and at worst outright misrepresent, the complex modalities and temporal rhythms of vegetal life. To fully exploit the inherent vulnerabilities of modern bioenergy as a governance logic, then, demands a return—I argue—to the literal roots of that logic, found in plants themselves. In short, it demands that we attune ourselves more consciously and deliberately to the distinctiveness and "never-fully-predictable alterity" of the vegetal kingdom, rather than seeing in plants a straightforward, natural embodiment of the comfortably familiar, productivist energy cultures to which the fossil fuel era has given rise.[28] The question of how this difficult shift might be achieved in practice is the focus of the next section.

Turning over a new leaf: post-productive plant potentials

Since the early 2010s, Western philosophy and social theory have been swept by a "plant turn", one embracing the vegetal kingdom not just as symbolically and materially constitutive of human cultures, but as a radically distinctive, autonomous sphere of life "at the fulcrum of its world".[29] For many key thinkers at the centre of this turn, plants have unfairly been allocated a subordinate status by almost all Western thinking on the nature of life, occupying what might be termed "the margin of the margin".[30] As Jeffrey Nealon points out, this is a script which conventionally casts vegetal life as inferior not just to humans but also to animals.[31] Burgeoning developments in certain, somewhat heterodox areas of plant science, concerned among other things with the potential existence of plant memory, intelligence or indeed communication, have offered significant fuel for some of these philosophical contributions.[32] These same developments, however, have also sometimes reinforced a propensity within Western science of attempting to identify in plants anatomical features deemed "analogous" to those already known to play vital roles in human or animal organisms.[33]

In seeking to properly understand plants, moreover, these contributions can sometimes ironically expunge the immense heterogeneity of the vegetal kingdom, in favour of a more reductive view of plants as a "uniform green mass".[34] According to some critical commentators, therefore, far from building genuinely new understandings of plants on their own terms,[35] these accounts more often succumb to "an injurious and narcissistic anthropomorphism".[36] Keeping a watchful eye out for the stealthy creep of anthropomorphism, however, does nothing to negate an even more fundamental problem with many contributions to the so-called "plant turn", found in the frequent bracketing out of rich strands of thinking about plant life to be found beyond the confines of Western knowledge frameworks, whether scientific, philosophical or otherwise. Typically, these are accounts emerging from neither cerebral reflection nor laboratory experimentation, but rather from the lived experience of close, intimate relationships with plants cultivated over generations by diverse indigenous communities, frequently foregrounding notions of care, conviviality, kinship and mutual flourishing.[37] If the challenge of coming to know plants on their own terms, of developing a truly authentic "ontophytology",[38] is one inherently fraught with methodological and ethical dilemmas, indigenous accounts would ultimately suggest that this arises less from our status as human beings than it does from the fact that lifeforms of all kinds are not pregiven, but rather embroiled in a continuous process of "becoming" through messy, situated and ongoing entanglements with one another.[39]

Against the backdrop formed by the wider "plant turn" and its critics, the idea of vegetal labour which I have advocated in this book might at first seem no less problematic than other concepts ostensibly "projecting" human characteristics onto the vegetal world. Yet the contention that plants can labour is not intended to serve as a philosophical assertion about the intrinsic nature of plant life, so much as a political provocation. To recognise and name plants as vegetal labourers offers, I have argued, a fecund starting point from which to contest the historically specific arrangements within which diverse forms of plant life are apprehended, brought to heel and enrolled within processes of economic value production—certainly more so than well-established but highly instrumental concepts like "natural capital" or "ecosystem services". Two further ideas about vegetal labour also bear restating at this juncture. Firstly, the concept insists on a view of all labour as constituted through situated, metabolic relations, rather than as the exclusive achievement of individual subjects (whether human, planty or otherwise).[40] Vegetal labour, from this standpoint, does not have essential qualities which can be abstracted from the place-specific, collaborative working arrangements within which particular species of plants are enrolled to achieve specific end goals. Secondly and no less importantly, vegetal labour, like all labour, is never an end in itself, but a means to an end—it can only ever form "a partial descriptor of lives, relationships, and activities", rather than offering a justification for their existence.[41] To attune ourselves more closely to plant alterity, in light of these two points, requires in the first instance a commitment to thinking more creatively about what vegetal labour could be for—to be open, in other words, to the possibilities of working with plants for a much more diverse range of purposes than those currently envisaged as virtuous and productive within the prevailing regulatory logic of modern bioenergy. But it also requires an acknowledgement that no particular manifestation of vegetal labour—regardless of the kinds of relationships between humans, plants and other lifeforms on which it relies—can ever claim to be properly faithful to the "true" nature of plants themselves. Contesting the ways in which diverse plants are currently enrolled as labourers, in other words, is not meant to help uncover the fundamental secrets of vegetal life, or even to isolate the ultimate "species being" of plants.[42] Instead, I am simply calling for a more modest commitment to asking what *else* plants might be able to teach us about ourselves, and especially about our relationships with energy, in an era otherwise dangerously devoid of concepts capable of disrupting the productivist and extractivist logics of the fossil energy era.[43]

This book does not claim to offer an exhaustive account of all of the possible ways in which today's prevailing regulatory logic of modern bioenergy might be transformed, or at least resisted, through more conscious efforts to notice the diverse capacities and potentials of the vegetal kingdom. Nonetheless, by drawing on examples of plants already widely enrolled into modern bioenergy production, I do aim to outline—even if only in provisional terms—some aspects of vegetal difference that might help us to rethink the role and purpose of energy in a post-fossil fuel age. At the end of Chapter 1, I suggested that plants might have teachings to convey to human societies about the availability and immediacy of diverse energy resources, about energy's links to productivity and growth and finally about the nature of resource scarcity and waste. With respect firstly to questions of availability and immediacy, a crucial tension exists in mainstream energy policy discourse, between accounts of bioenergy as inherently more reliable than other, supposedly more intermittent forms of renewable energy generation on the one hand, and on the other hand the simple fact that plants do not grow instantaneously. On a planetary scale, what Andreas Malm describes as flow-based renewables—chiefly wind, solar and hydropower—are so abundantly and continuously available as to render all current human demands for energy vanishingly small.[44] Intermittency in the actual supply of these forms of energy is undeniable, but this arises not from the nature of the resources themselves—all of which, in a warming climate, will only become more ubiquitous, not less—so much as deficiencies in the raw density and geographical interconnectivity of the physical infrastructures which human societies have (to date at least) constructed to harness them. In direct contrast to "flow" energy resources, bioenergy's availability in the ambient environment is anything but immediate. The most prominent feedstock used to make combustible wood pellets for the UK electricity sector, the loblolly pine (*Pinus taeda*), for example, despite having been optimised into a very fast-growing species through decades of scientific research and commercial investment, still ultimately takes at least ten years to reach a size deemed marketable for conversion into exportable biomass fuel.[45] Similarly, the derivation of bioethanol from various cultivars of sugarcane (*Saccharum* genus), which have long served as large-scale replacements for fossil-based petrol in Brazil and now increasingly help to propel cars driven by citizens in the US, Europe and elsewhere,[46] cannot short-circuit this plant's typically year-long growing cycle.

Of course, even as the exigencies of what has been termed "biological time"[47] remain a stubborn impediment to the instantaneous creation of biomass feedstocks, mainstream understandings of modern bioenergy overlook both the time and the vegetal labour required to produce these resources, preferring instead to foreground the material affinities between biofuels and fossil fuels at the point of their consumption. To wit, in the end, both kinds of fuel are valued for their status as combustible hydrocarbons which can be stockpiled and deployed on demand, and which therefore promise to sustain intensive energy consumption as a continuous, unbroken affair, rather than one susceptible to involuntary contractions or, worse still, sudden "blackouts".[48] A narrow focus on finished fuel products, however, elides the wide array of temporalities and socioecological contexts upon and within which plants actually do photosynthesise and grow—whether that be in commercially managed forests of the US South, wheat fields of rural England or palm oil plantations of Malaysia or Indonesia. Paying closer attention to the circumstances of bioenergy's making, I argue, would force us to notice and appreciate the time-consuming metabolic work which plants must undertake to reassemble solar radiation, water and carbon dioxide into complex organic compounds. From such a standpoint, bioenergy would be revealed as a resource neither strictly exhaustible nor immediately replaceable—no matter how much investment and effort are directed towards their "industrial improvement", plants will always "make us wait" for fuel itself.[49] Through a closer attunement, then, not just to the innate inexhaustibility of plant growth, but also to its distinct resource temporalities when compared to alternative renewables, we might perhaps come to appreciate fuels themselves as a reward needing to be earned through patience, rather than as a pregiven "gift" of nature to be accumulated, hoarded and thence consumed on a continuous, uninterrupted and ultimately unthinking basis. Crucially, this idea does not amount to a call for a generalised slowing down in all rates of energy consumption—certainly the point here is not to invoke "a naive longing for an invented, utopic, slow-moving society".[50] Rather, I am calling for a critical reinterrogation of commitments to the idea that energy consumption should *always* be as ceaseless and relentless as possible. The point is to reflect, in other words, on the possibility of more undulating, punctuated cadences of resource use centred around periods of deliberate abstinence, reflection and restraint on the one hand, and—on the other—around altogether more cathartic episodes of squander and excess.[51]

As we wait for the clock of solar time to be gradually "wound up" by plants,[52] we might reflect not merely on the commitments of dominant energy cultures of the global North to consumption without pause, but also towards continuous expansion and growth. Plants may be easily upheld as natural embodiments of productivity and growth, but in reality the temporalities of vegetal metabolism are complexly non-linear; certainly plants do not always aim "to work harder, faster, and better".[53] Left undisturbed, for example, loblolly pine's prodigious juvenile growth stage soon gives way to a more measured period of only modest expansion, tending ultimately towards stasis—while 25-year-old loblolly pines typically measure 30 metres tall, an individual ten times as old will rarely exceed a height of 50 metres.[54] In a very similar vein, meanwhile, Kate Neville has recently written of the evergreen subalpine fir tree that its seasonal growth "cycles ... reject dichotomies, allowing for speed and slowness, work and idleness".[55] The same point applies to the growth temporalities exhibited by a vast array of crop plants—such as sugarcane or corn—used to generate biofuel feedstocks, however ostensibly revered these might be for their high yields come harvest time. More fundamentally, and as is hinted at by growing efforts in plant biotechnology to fix so-called "glitches" inherent to the process, even photosynthesis itself poses challenges to productivist thinking.[56] The simple fact that leaves are green, for example, confirms that plants do not make use of all of the radiometric energy which could theoretically be harnessed for photosynthesis. Instead, this process is driven principally by red and blue light, leaving untapped most of the potential energy which plants could theoretically direct towards maximising their own growth.[57] Finally, and perhaps even more straightforwardly, it must also be reiterated that plants ultimately appear as useful energy resources to humans at least as much because of their capacities to bring energy to rest as because of their capacities to put energy to work. The formation by plants of organic matter, especially in the forms of lignocellulosic biomass and wood, offers a lesson in growth, no doubt—but it also constitutes a study in the patient, painstaking conversion of restless circulations of light, carbon dioxide and water into something remarkably stable, robust and enduring. The implicit invitation, perhaps, is then for us to abandon capitalism's pursuit—in Giorgos Kallis's terms—of the "permanent transgression of a constantly expanding frontier",[58] and instead to regard stasis and stability—or even idleness—as equally precious states to those of dynamism, production and growth.

When it comes to understanding the relationships between growth and life itself, meanwhile, plants may have even more radical lessons to pass on. For plant philosopher Michael Marder, for instance, vegetal growth cannot be seen as a "purely quantitative increase in extension", nor as the inexorable realisation of a final form which was already genetically pre-scripted into the seed itself—rather, it is something altogether more "unpredictable and eventful", not least due to its being inextricably open to the influence of a wider environmental milieu.[59] In subscribing to this view, Marder finds himself in agreement with Luce Irigaray, in her similar view of the growth of plants—indeed, the very being of the vegetal kingdom—as "without any object" or final overriding purpose.[60] From the vantage point of today's prevailing modern bioenergy logic, these contentions might seem to miss the point; here the objective of cultivating plants is self-evidently to serve as feedstocks for a more sustainable form of energy generation than has historically been provided by fossil fuels. Scientific and commercial initiatives aiming to make loblolly pines grow more quickly, however, or to genetically engineer crops like *Miscanthus* and sorghum to help them more successfully emulate oil itself, are also projects which seek to compress whole stages of cultivation normally predicated on patience and "the capacity to wait ... for the time the plants themselves require to mature".[61] In so doing, these projects squeeze out any potential for humans not just to notice and respect the time of plants, but also to recognise and embrace the inherent open-endedness of what plants themselves might be able to become. Plants are, in short, "de-eventalized" by the techno-capitalist arena of modern bioenergy, to borrow the terms used by Marder in the first epigraph to this chapter. Moreover, when plants are upheld as paragons of productivism and continuous expansion precisely in order to legitimise an economy governed by the same ideals, this "de-eventalization" serves to deny not simply the open-endedness of plants, but the open-endedness of human relationships with energy too.

Finally, to scarcity and waste. On the surface, of course, plants clearly do offer the promise of transcending the apparent limits to economic growth imposed by finite fossil fuel resources—as indicated by the near-axiomatic links drawn in policy discourse between modern bioenergy expansion and enhanced energy security. In offering the prospect, as Denis Hayes put it in the late 1970s, of "dramatically shortening the time geological forces take to convert vegetation into oil, gas and coal",[62] plants might even be seen as saviours in the struggle to resist entropy itself. From this standpoint, the beguiling "agencies of rearrangement" exhibited by plants,[63] and especially their abilities to fashion complex organic compounds apparently out of nothing in a matter of just years or even months, short-circuit the deep-time temporalities of fossil fuel formation to such a stunning degree that they could almost be said to defy the principles of thermodynamics. Yet the problem with cultural fears about entropy is that they have been historically stoked not just by scientific claims about thermodynamic "laws", but also by wider protagonists of nineteenth-century European industrialisation and colonial expansion—projects both undergirded by the frenzied exploitation of fossil fuels.[64] Whenever energy is "lost" or "wasted", these aberrations can only properly be defined in relation to specific ideas about what makes energy useful or purposeful to begin with. All too often, at least in the Western imaginations consumed by a "bias towards dynamism over stasis", the fundamental point of energy has been presumed as helping society to achieve "constant motion, action, dynamism, growth—with progressive effects assumed to follow post-hoc".[65] Energy use that does not maximise work and productivity, in short, represents scandalous waste. But, even as they help to drastically compress conventional temporalities of hydrocarbon resource-making when compared to coal, oil or gas, does that really mean that plants reinforce these understandings of work and waste by their very nature?

To return to Michael Marder once more, all plants ultimately perform a kind of "ontological indifference" to their existence, and certainly to the particular ends to which they are put by human communities.[66] The loblolly pine does not care whether it is processed into wood pellets after ten years of rapid growth, or instead left to grow more languidly towards an age of one hundred years before being put to the same use. Sugarcane, likewise, does not condemn the agricultural regime which burns bagasse—the inedible, fibrous biomass left over following sugar extraction—in the open air, or which feeds this matter directly to livestock, rather than seeking to convert it into "advanced" cellulosic biofuel. Through their indifference to the logics not just of work and waste, but arguably also of self-preservation, plants exhibit relationships with energy entirely devoid of "a calculation, a planning, [or] a goal orientation".[67] In this way, the vegetal kingdom offers us a direct connection to an energy culture which transcends the narrow, utilitarian terms of what Georges Bataille termed the "particular" or "rational economy",[68] opening an aperture instead for the embrace—even if only on an infrequent, fleeting basis—of a mode of energy expenditure motivated precisely by the desire to squander; to engage, in other words, in a "blind spending of the intimate world".[69] To be clear, my aim here is not to usher in or condone the unhindered expansion of modern bioenergy logics, to encompass effectively limitless biomass burning. Rather, the point is to suggest that a genuinely vegetal energy culture might reasonably exercise patience and parsimony in its use of fuel resources not to secure ongoing societal compliance with any calculative logic of global environmental sustainability, or even plain old resource efficiency, but rather in order to allow for the concepts of efficiency, usefulness and waste themselves to be periodically rejected in favour of more concentrated outbursts of decadent excess. After all, as Bataille puts it, "[O]n the surface of the globe, for living matter in general, energy is always in excess ... The choice is limited to how the wealth is to be squandered".[70]

Dormant seeds? Putting down roots for alternative energy futures

Several key characteristics of the vegetal world—its theoretical inexhaustibility, the non-linear temporalities and open-ended pathways of its growth, its tendency towards rest as much as to exertion, and its ultimate indifference to the notions of use—offer political potentials quite unlike those afforded by other major energy resources available to human societies today. Plants, in other words, are not just living feedstocks whose metabolic work promises to facilitate more sustainable forms of hydrocarbon production. Rather, they constitute a distinct material and ecological foundation upon which to reshape energy–society relationships inherited from the fossil fuel era—relationships linking both individual identities and broader societal ideals to the ever-increasing generation and consumption of energy. More specifically, I have argued that plants can and should be seen as inviting us to rethink several fundamental features of our prevailing energy cultures, as follows. Firstly, an attunement to the time plants take to grow should encourage us to adopt more patient and respectful attitudes towards the making of fuels, and indeed to revisit the cadence of our relationships with energy resources of all kinds. Secondly, the non-linearity of plant growth should prompt us to temper the reification of continuous (and continuously expanding) production and consumption, arguably inconsistent with the ideal of a "sustainable" or "circular" economy in any case,[71] with an equal embrace of stability, rest and perhaps even idleness. Thirdly, when understood as inherently unpredictable and eventful, rather than as a formal process of pre-scripted expansion, plant growth should give us pause to reflect anew on the possibility of cultivating relationships with energy whose purpose is itself open-ended, and not simply presumed a priori as being oriented towards growth for the sake of growth. Fourthly and finally, through their indifference to notions not merely of usefulness but also of self-preservation and "self-maintenance",[72] plants should—I argue—incite us finally to reject nineteenth-century European imperatives to maximise work and minimise waste, as well as their attendant calculative logics of efficiency and sustainability, in favour of more occasional, cathartic outbursts of deliberate squander and excess.

Calls issued by Western academics to "ally ourselves more closely with plants",[73] of course, have long been pre-empted within an array of non-Western ontologies, where views of plants as "teachers" or "kin" animate multispecies relationships predicated not upon perpetual expansion and growth, but mutual flourishing.[74] Accordingly, in concluding this book, any lessons that I have suggested plants might be able to teach us about energy are intended to be neither exhaustive nor tightly prescriptive. Nonetheless, whatever their precise ethos, it is difficult to imagine that a closer attunement to the alterity and political affordances of plants as energy resources can ever be cultivated in a world where massive quantities of vegetal matter are routinely uprooted and appropriated, often from far-flung locations, simply to help "decarbonise" the high-consumption energy systems which predominantly serve existing global centres of economic wealth and power.[75] Instead, if meaningful alternatives to today's large-scale, industrial modern bioenergy regimes are to be created, this will most likely occur in settings where the terms and end goals of bioenergy production have been locally self-determined rather than externally imposed. After all, if plants do indeed present vital opportunities to reflect anew on the roles that energy should play in shaping societies and economies in a supposedly post-fossil fuel era, it is no less crucial to reflect on who should supply the answers to such existential questions on the ground. Why should commercial operations in working forests in Mississippi or Louisiana be driven, for example, by foreign interests seeking to maximise wood output for the production of supposedly "carbon-neutral" electricity abroad, instead of by local communities wishing to power activities designed to meet their own, self-determined standards of functionality, prosperity or even simple enjoyment? Similarly, why shouldn't agricultural landscapes of the American Midwest play host, over the coming decades, to the cultivation of modest volumes of sugarcane or *Miscanthus* (or indeed other crops altogether) for use in local heat and power generation, rather than being appropriated to grow engineered "oilcane" destined for conversion into jet fuel propelling airliners through the skies above those very same landscapes?

In raising these questions, the final aim of this book is not to imply the innate superiority of any particular mode of bioenergy production above all others. These closing remarks should not be read, for example, as a blanket endorsement of "mundane" forms of bioenergy over "modern" bioenergy,[76] nor necessarily even as anti-(bio)technological. Nonetheless, there remains a pressing need to avoid simply accepting accounts of productivism as inherently virtuous; to pause and actively ask, as Kate Neville has recently done, "[W]hose labours, and which labours, are barreling ahead, and which ones move—or could move—at a different tempo?"[77] With these points in mind, my suggestion here is simply that the characteristics of plants present society—whether we should choose to recognise the fact or not—with choices about the roles that energy could and should play in shaping future forms of social and economic life. We could, of course, ignore these choices and continue on our current path of trying to squeeze as much power from plants as possible, placing our faith in the ostensibly indefatigable capacities of the vegetal world to help realise fantasies of a "good" bioeconomy that is at once apparently sustainable and circular, and yet simultaneously also characterised by continuous expansion and growth.[78] Alternatively, we could instead work with plants to cultivate myriad alternative relationships with resources—and indeed resource limits—animated not by any thoughtless drive for continuously expanding consumption, but rather by the deliberate and thoughtful interrogation of energy's "expansive, multi-dimensional" potentials[79] to create quite different human worlds, far less preoccupied with pursuing growth for the sake of growth. In short, what is being advanced is a call for greater protections to be accorded to the basic autonomy that will be necessary for diverse, self-determined forms of place-based collaboration between humans and plants to come into being and flourish, with as yet indeterminate implications for the kinds of energy cultures that might result.[80] While the seeds of such cultures may be dormant for now, their germination depends ultimately upon emancipating diverse agricultural and forestry landscapes from the potent colonising forces of today's modern "bioenergy boom",[81] a boom which, while global in ambition, serves in practice the interests only of those nations already advantaged by centuries of fossil fuel-powered, European industrial and colonial conquest. By releasing such landscapes from the suffocating logic of productivism, genuinely alternative ideas about energy's purpose might yet, with a little help from plants themselves, begin to lay down new roots.

Notes

1 Daggett, "Energy and domination."
2 Ernwein *et al.*, *The Work That Plants Do*; Palmer, "Rethinking plant power"; and Lohmann, "Bioenergy, thermodynamics and inequalities."
3 Foucault, "The confession of the flesh," 194.
4 Energy Institute, *Statistical Review of World Energy 2024*, 65; and International Energy Agency, *Renewables 2023*, 94.
5 Proskurina *et al.*, "Global biomass trade for energy."
6 Fressoz, *More and More and More*, 126.

7 Aykut *et al.*, "'Incantatory' governance"; and Rayner, "What might Evans-Pritchard have made of two degrees?"
8 Daggett, *The Birth of Energy*, 27; and Sieferle, *The Subterranean Forest.*
9 Bakker and Bridge, "Material worlds redux," 47.
10 Boyd *et al.*, "Industrial dynamics and the problem of nature."
11 Malm and Carton, *Overshoot*, 63.
12 Giampietro, "On the circular bioeconomy and decoupling," 143. As Zwier and colleagues point out, a "truly circular economy would render terms like input and output redundant" (Zwier *et al.*, "The idea of a zero-waste humanity," 355).
13 Zwier *et al.*, "The ideal of a zero-waste humanity." As Zwier and colleagues admit, a "zero-waste" economy may well be impossible to achieve in reality, given the laws of thermodynamics. As a thought experiment, however, their argument serves to illustrate fundamental contradictions between commitments to circularity and growth.
14 Zwier *et al.*, "The ideal of a zero-waste humanity," 367.
15 Palmer and Carton, "Carbon removal as carbon revival?"
16 Barua, "Nonhuman labor, encounter value, spectacular accumulation," 283.
17 Juskus, "Sacrifice zones."
18 Mol, "Boundless biofuels?"
19 The White House, "President signs Energy Policy Act"; and US Department of Energy, *2023 Billion-Ton Report.*
20 Baka and Bailis, "Wasteland energy-scapes"; and Chaudhary, "The extractive circuit," 32.
21 International Energy Agency, "Modern bioenergy leads the growth of all renewables to 2023."
22 Harlan and Baka, "Stacked energyscapes," 2.
23 Fressoz, *More and More and More*, 9. See also Cointe, "Decarbonation as substitution."
24 See Goldstein, "Net zero and settler futurity."
25 Carton, "'Fixing' climate change by mortgaging the future"; see also Malm and Carton, *Overshoot.*
26 Goldstein, "Net zero and settler futurity," 31.
27 Prudham, *Knock on Wood*, 8.
28 Ginn and Connor, "Vegetal hydropoetics," 497.
29 Marder, *Plant-Thinking*, 8.
30 Marder, *Plant-Thinking*, 2.
31 Nealon, *Plant Theory*, 36.
32 See Pollan, "The intelligent plant"; and Mancuso, *The Revolutionary Genius of Plants.*
33 Coccia, *The Life of Plants*, 126.
34 Ernwein *et al.*, "Introduction: The work that plants do," 18.
35 For Michael Marder, "[W]e are sure not to hear plants speak if we continue to hold onto the anthropocentric prejudice that sees in our intelligence, cognition, and languages the gold standards of intelligence, cognition, and language as such" (Marder, "To hear plants speak," 104).
36 Lawrence, "Listening to plants," 635.

37 See for example Kimmerer, *Braiding Sweetgrass*; and Kohn, *How Forests Think.*
38 Marder, *Plant-Thinking*, 94.
39 Irigaray, "What the vegetal world says to us."
40 Robertson and Wainwright, "The value of nature to the state," 895; and Smith, *Uneven Development*, 53.
41 Battistoni, "Bringing in the work of nature," 24.
42 Fair and McMullen, "Toward a theory of nonhuman species-being."
43 In referring to plants explicitly as potential "teachers", I here deliberately invoke a mode of plant–human relations favoured by Robin Wall Kimmerer and Luce Irigaray (Kimmerer, *Braiding Sweetgrass*; and Irigaray, "What the vegetal world says to us").
44 According to Malm, "[T]he rate at which the earth intercepts sunlight is nearly 10,000 times greater than the entire energy flux humans currently muster" (Malm, *Fossil Capital*, 367).
45 Palmer, "Forest fuels," 168.
46 Eaglin, *Sweet Fuel.*
47 Prudham, "Taming trees"; and Adam, *Timescapes of Modernity.*
48 Walker, "Keeping the lights on."
49 Sedgewick, *Coffeeland*, 85.
50 Neville, *Going to Seed*, 77.
51 Bataille, *The Accursed Share.*
52 Tyndall, *Heat Considered as a Mode of Motion*, 507.
53 Boyd *et al.*, "Industrial dynamics and the problem of nature," 564.
54 Palmer, "Forest fuels."
55 Neville, *Going to Seed*, 77.
56 Realizing Increased Photosynthetic Efficiency, "Scientists engineer shortcut for photosynthetic glitch, boost crop growth by 40 percent."
57 See Nansen, "Unraveling the 'radiometric signals' from green leaves."
58 Kallis, *Limits*, 115.
59 Marder, "The sense of seeds, or seminal events," 88–89.
60 Irigaray, "What the vegetal world says to us," 134.
61 Marder, "The sense of seeds, or seminal events," 93.
62 Hayes, "Plant power," 349.
63 Myers, "Photosynthetic mattering," 125.
64 Daggett, *The Birth of Energy.*
65 Daggett, *The Birth of Energy*, 18.
66 Marder, *Plant-Thinking*, 134.
67 Stoekl, *Bataille's Peak*, 143.
68 Bataille, *The Accursed Share.*
69 Stoekl, *Bataille's Peak*, 144.
70 Bataille, *The Accursed Share*, 23.
71 Zwier *et al.*, "The ideal of a zero-waste humanity."
72 Zwier and Blok, "Energetic ethics."
73 Lawrence, "Listening to plants," 634.
74 Kimmerer, *Braiding Sweetgrass*; Turner *et al.*, "Well grounded"; and Hitch and Grignon, "A forest of energy."

75 Palmer and Carton, "Carbon removal as carbon revival?" 6.
76 Chatti *et al.*, "Exploring the mundane."
77 Neville, *Going to Seed*, 77.
78 Asdal *et al.*, " 'The good economy' "; and Zwier *et al.*, "The ideal of a zero-waste humanity."
79 Daggett, *The Birth of Energy*, 4.
80 On autonomy, see Collard *et al.*, "A manifesto for abundant futures."
81 Lohmann, "Bioenergy, thermodynamics and inequalities."

Bibliography

Adam, Barbara. *Timescapes of Modernity: The Environment and Invisible Hazards*. London: Routledge, 1998

Adam, David. "Kelly orders biofuels review." *The Guardian*, 21 February 2008. https://www.theguardian.com/environment/2008/feb/21/biofuels.transport?gusrc=rss (accessed 22 October 2024)

Allison, Graham. *Essence of Decision: Explaining the Cuban Missile Crisis*. Boston, MA: Little, Brown and Company, 1971

Alonso-Fradejas, Alberto. "'Leaving no one unscathed' in sustainability transitions: The life purging agro-extractivism of corporate renewables." *Journal of Rural Studies* 81 (2021): 127–138

Ambrose, Jillian, and Niko Kommenda. "Britain breaks record for coal-free power generation." *The Guardian*, 28 April 2020. https://www.theguardian.com/business/2020/apr/28/britain-breaks-record-for-coal-free-power-generation (accessed 4 February 2025)

Anderson, Kevin, and Glen Peters. "The trouble with negative emissions." *Science* 354, no. 6309 (2016): 182–183

Anderton, Karen, and James Palmer. "Evidence-based policy as iterative learning: The case of EU biofuels targets." *Contemporary Social Science* 10, no. 2 (2015): 138–147

Apostolopoulou, Evangelia. *Nature Swapped and Nature Lost: Biodiversity Offsetting, Urbanization and Social Justice*. Basingstoke: Palgrave Macmillan, 2020

Asdal, Kristin, Béatrice Cointe, Bård Hobæk *et al.* "'The good economy': A conceptual and empirical move for investigating how economies and the good are entangled." *BioSocieties* 18, no. 1 (2023): 1–24

Aykut, Stefan C., Edouard Morena, and Jean Foyer. "'Incantatory' governance: Global climate politics' performative turn and its wider significance for global politics." *International Politics* 58 (2020): 519–540

Bäckstrand, Karin, James Meadowcroft, and Michael Oppenheimer. "The politics and policy of carbon capture and storage: Framing an emergent technology." *Global Environmental Change* 21, no. 2 (2011): 275–281

Baka, Jennifer. "Making space for energy: Wasteland development, enclosures, and energy dispossessions." *Antipode* 49, no. 4 (2017): 977–996

Baka, Jennifer, and Robert Bailis. "Wasteland energy-scapes: A comparative energy flow analysis of India's biofuel and biomass economies." *Ecological Economics* 108 (2014): 8–17

Baker, J.B., and G. Langdon. "*Pinus taeda*." In *Silvics of North America: 1. Conifers; 2. Hardwoods, Agriculture Handbook 654*, edited by R.M. Burns and B.H.

Honkala, 497–512. Washington, DC: United States Department of Agriculture Forest Service, 1990

Bakker, Karen. *An Uncooperative Commodity: Privatizing Water in England and Wales*. Oxford: Oxford University Press, 2003

Bakker, Karen. "Neoliberal nature, ecological fixes, and the pitfalls of comparative research." *Environment and Planning A* 41 (2009): 1781–1787

Bakker, Karen, and Gavin Bridge. "Material worlds? Resource geographies and the 'matter of nature'." *Progress in Human Geography* 30, no. 1 (2006): 5–27

Bakker, Karen, and Gavin Bridge. "Materials worlds redux: Mobilizing materiality within critical resource geography." In *The Routledge Handbook of Critical Resource Geography*, edited by Matthew Himley, Elizabeth Havice, and Gabriela Valdivia, 43–56. London: Routledge, 2021

Barca, Stefania. "Labour and the ecological crisis: The eco-modernist dilemma in western Marxism(s) (1970s–2000s)." *Geoforum* 98 (2019): 226–235

Barua, Maan. "Nonhuman labour, encounter value, spectacular accumulation: The geographies of a lively commodity." *Transactions of the Institute of British Geographers* 42, no. 2 (2017): 274–288

Barua, Maan. "Animating capital: Work, commodities, circulation." *Progress in Human Geography* 43, no. 4 (2019): 650–669

Barua, Maan. "Affective economies, pandas, and the atmospheric politics of lively capital." *Transactions of the Institute of British Geographers* 45, no. 3 (2020): 678–692

Bataille, Georges. *The Accursed Share: An Essay on General Economy*, vol. I. Brooklyn, NY: Zone Books, 1988[1949]

Battistoni, Alyssa. "Bringing in the work of nature: From natural capital to hybrid labor." *Political Theory* 45, no. 1 (2017): 5–31

Battistoni, Alyssa. *Free Gifts: Capitalism and the Politics of Nature*. Princeton, NJ: Princeton University Press, 2025

BBC News. "Green fuel obligation announced." British Broadcasting Corporation, 10 November 2005. http://news.bbc.co.uk/1/hi/uk/4424990.stm (accessed 15 September 2024)

BBC News. "PM opens major biofuels complex." British Broadcasting Corporation, 30 June 2006. http://news.bbc.co.uk/1/hi/england/tees/5133590.stm (accessed 19 September 2024)

BBC News. "Biofuels 'crime against humanity'." British Broadcasting Corporation, 27 October 2007. http://news.bbc.co.uk/1/hi/7065061.stm (accessed 4 September 2024)

BBC News. "EU rethinks biofuels guidelines." British Broadcasting Corporation, 14 January 2008. http://news.bbc.co.uk/1/hi/7186380.stm (accessed 13 February 2025)

BBC News. "Call for delay to biofuels policy." British Broadcasting Corporation, 24 March 2008. http://news.bbc.co.uk/1/hi/sci/tech/7309099.stm (accessed 13 February 2025)

BBC News. "Britain goes coal free as renewables edge out fossil fuels." British Broadcasting Corporation, 10 June 2020. https://www.bbc.co.uk/news/science-environment-52973089 (accessed 17 March 2025)

BBC News. "Groundbreaking transatlantic flight using greener fuel lands in the US." British Broadcasting Corporation, 28 November 2023. https://www.bbc.co.uk/news/business-67548961 (accessed 30 January 2025)

Bearak, Max, Dionne Searcey, and Mira Rojanasakul. "Airlines race toward a future of powering their jets with corn." *The New York Times*, 30 November 2023. https://www.nytimes.com/interactive/2023/11/30/climate/airlines-jet-fuel-ethanol-corn.html (accessed 31 January 2025)

Beck, Silke, and Martin Mahony. "The politics of anticipation: The IPCC and the negative emissions technologies experience." *Global Sustainability* 1, no. e8 (2018): 1–8

Beddington, John, Steven Berry, Ken Caldeira *et al.* "Letter from scientists to the EU Parliament regarding forest biomass." 9 January 2018. https://www.euractiv.com/wp-content/uploads/sites/2/2018/01/Letter-of-Scientists-on-Use-of-Forest-Biomass-for-Bioenergy-January-12–2018.pdf (accessed 18 March 2025)

Beldo, Les. "Metabolic labor: Broiler chickens and the exploitation of vitality." *Environmental Humanities* 9, no. 1 (2017): 108–128

Bellamy, Rob, Mathias Fridahl, Javier Lezaun *et al.* "Incentivising bioenergy with carbon capture and storage (BECCS) responsibly: Comparing stakeholder policy preferences in the United Kingdom and Sweden." *Environmental Science and Policy* 116 (2021): 47–55

Bellamy, Rob, Javier Lezaun, and James Palmer. "Perceptions of bioenergy with carbon capture and storage in different policy scenarios." *Nature Communications* 10, no. 743 (2019): 1–9

Bellamy, Rob, and James Palmer. "Geoengineering and geographers: Rewriting the Earth in what image?" *Area* 51, no. 3 (2019): 524–531

Besky, Sarah, and Alex Blanchette, eds, *How Nature Works: Rethinking Labor on a Troubled Planet*. Albuquerque: University of New Mexico Press, 2019

Birch, Kean. *Neoliberal Bio-economies: The Co-construction of Markets and Natures*. Basingstoke: Palgrave Macmillan, 2018

Birch, Kean, and Kirby Calvert. "Rethinking 'drop-in' biofuels: On the political materialities of bioenergy." *Science and Technology Studies* 28, no. 1 (2015): 52–72

Birch, Kean, and David Tyfield. "Theorizing the bioeconomy: Biovalue, biocapital, bioeconomics or ... what?" *Science, Technology, & Human Values* 38, no. 3 (2013): 299–327

Bonneuil, Christophe, and Jean-Basptise Fressoz. *The Shock of the Anthropocene: The Earth, History and Us*. London: Verso, 2017

Borras, Saturnino M., Jr., Philip McMichael, and Ian Scoones. "The politics of biofuels, land and agrarian change: Editors' introduction." *Journal of Peasant Studies* 37, no. 4 (2010): 575–592

Boyd, William. *The Slain Wood: Papermaking and Its Environmental Consequences in the American South*. Baltimore, MD: Johns Hopkins University Press, 2015

Boyd, William, W. Scott Prudham, and Rachel A. Schurman. "Industrial dynamics and the problem of nature." *Society and Natural Resources* 14, no. 7 (2001): 555–570

Boyer, Dominic. "Energopolitics and the anthropology of energy." *Anthropology News* 52, no. 5 (May 2011): 5–7

Boyer, Dominic. *Energopolitics: Wind and Power in the Anthropocene*. Durham, NC: Duke University Press, 2019

Boyer, Miriam, Franziska Kusche, Sarah Hackfort *et al.* "The making of sustainability: Ideological strategies, the materiality of nature, and biomass use in the bioeconomy." *Sustainability Science* 18 (2023): 675–688

Braun, Bruce. "Environmental issues: Inventive life." *Progress in Human Geography* 32, no. 5 (2008): 667–679

Brice, Jeremy. "Attending to grape vines: Perceptual practices, planty agencies and multiple temporalities in Australian viticulture." *Social & Cultural Geography* 15, no. 8 (2014): 942–965

Bridge, Gavin. "Resource triumphalism: Postindustrial narratives of primary commodity production." *Environment and Planning A* 33, no. 12 (2001): 2149–2173

Bridge, Gavin. "Resource geographies II: The resource-state nexus." *Progress in Human Geography* 38, no. 1 (2014): 118–130

Bridge, Gavin, Stefan Bouzarovski, Michael Bradshaw, and Nick Eyre. "Geographies of energy transition: Space, place and the low-carbon economy." *Energy Policy* 53 (2013): 331–340

Brock, Andrea, Benjamin K. Sovacool, and Andrew Hook. "Volatile photovoltaics: Green industrialization, sacrifice zones, and the political ecology of solar energy in Germany." *Annals of the American Association of Geographers* 111, no. 6 (2021): 1756–1778

Brown, Benjamin, and Sam Spiegel. "Coal, climate justice, and the cultural politics of energy transition." *Global Environmental Politics* 19, no. 2 (2019): 149–168

"Brown should stand firm on rising fuel prices." *The Guardian*, 4 June 2004. https://www.theguardian.com/politics/2004/jun/04/economy.oil (accessed 22 September 2024)

Buck, Holly J. *After Geoengineering: Climate Tragedy, Repair, and Restoration*. London: Verso, 2019

Burley, Helen, and Hannah Griffiths. "*Jatropha*: Wonder crop? Experience from Swaziland." *Friends of the Earth*, May 2009. https://www.foei.org/wp-content/uploads/2020/12/Jatropha-Swaziland.pdf (accessed 13 December 2025)

CABBI. *Filling the Gaps and Defining the Future of Bioenergy*. Urbana, IL: Centre for Advanced Bioenergy and Bioproducts Innovation, 2018

CABBI. "Building a better bioenergy crop: A Q&A with Erik Sacks." Centre for Advanced Bioenergy and Bioproducts Innovation, 6 April 2022. https://cabbi.bio/building-a-better-bioenergy-crop-a-qa-with-erik-sacks/ (accessed 5 February 2025)

CABBI. "Rubisco's role in global food and energy security." Centre for Advanced Bioenergy and Bioproducts Innovation, 17 December 2024. https://cabbi.bio/rubiscos-role-in-global-food-and-energy-security/ (accessed 5 February 2025)

CABBI. "From lab to field: CABBI pipeline delivers oil-rich sorghum." Centre for Advanced Bioenergy and Bioproducts Innovation, 17 January 2025. https://cabbi.bio/from-lab-to-field-cabbi-pipeline-delivers-oil-rich-sorghum/ (accessed 5 February 2025)

CABBI. "Feedstock production." Centre for Advanced Bioenergy and Bioproducts Innovation. https://cabbi.bio/research/feedstock-production/ (accessed 4 February 2025)

Carmody, Pádraig. *The New Scramble for Africa*. Cambridge: Polity, 2011

Carton, Wim. "'Fixing' climate change by mortgaging the future: Negative emissions, spatiotemporal fixes, and the political economy of delay." *Antipode* 51, no. 3 (2019): 750–769

Carton, Wim, Erik Jönsson, and Beatriz Bustos. "Revisiting the 'subsumption of nature': Resource use in times of environmental change." *Society & Natural Resources* 30, no. 7 (2017): 789–796

Castree, Noel. "Nature is dead! Long live nature!" *Environment and Planning A* 36, no. 2 (2004): 191–194

Castree, Noel. "Neoliberalising nature: The logics of deregulation and reregulation." *Environment and Planning A* 40, no. 1 (2008): 131–152

C-Capture. "£5M boost to scale up ground-breaking carbon capture pilot at Drax, UK's largest power station." June 2019. https://c-capture.co.uk/5m-boost-scale-ground-breaking-carbon-capture-pilot-drax-uks-largest-power-station/ (accessed 24 March 2025)

Charles, Michael B., Rachel Ryan, Neal Ryan, and Richard Oloruntoba. "Public policy and biofuels: The way forward?" *Energy Policy* 35 (2007): 5737–5746

Chatham House. *Woody Biomass for Power and Heat: Impacts on the Global Climate*. London: Royal Institute of International Affairs, 2017

Chatham House. *Net Zero and Beyond: What Role for Bioenergy with Carbon Capture and Storage*. London: Royal Institute of International Affairs, 2020

Chatti, Deepti, Matthew Archer, Myles Lennon, and Michael R. Dove. "Exploring the mundane: Towards an ethnographic approach to bioenergy." *Energy Research and Social Science* 30 (2017): 28–34

Chaudhary, Ajay Singh. "The extractive circuit: An exhausted planet at the end of growth." *The Baffler* 60 (2021): 26–38.

Chum, Helena, Andre Faaij, José Moreira *et al.* "Bioenergy." In *IPCC Special Report on Renewable Energy Sources and Climate Change Mitigation*, edited by Ottmar Edenhofer, Ramón Pichs-Madruga, Youba Sokona *et al.*, 209–331. Cambridge: Cambridge University Press, 2011

Clean Cooking Alliance. *The Value of Clean Cooking*. https://cleancooking.org/the-value-of-clean-cooking/ (accessed 21 February 2025)

Clean Skies for Tomorrow. "2030 ambition statement." World Economic Forum, 22 September 2021. https://www3.weforum.org/docs/WEF_EMBARGOED_CST_Ambition_Statement_for_Signatories.pdf (accessed 15 March 2025)

Climate Assembly UK. *The Path to Net Zero: Climate Assembly UK Full Report*. Climate Assembly UK, 2020

Coccia, Emanuele. *The Life of Plants: A Metaphysics of Mixture*. Medford, MA: Polity, 2019

Cockburn, H. "UK's record coal-free power run comes to an end." *Independent*, 17 June 2020. https://www.independent.co.uk/environment/coal-free-power-uk-record-time-2020-how-long-renewable-energy-a9570891.html (accessed 4 February 2025)

Cointe, Béatrice. "Decarbonation as substitution." *Cultural Anthropology Theorizing the Contemporary*, 12 November 2024. https://culanth.org/fieldsights/decarbonation-as-substitution (accessed 7 March 2025)

Cointe, Béatrice, and Hélène Guillemot. "A history of the 1.5°C target." *Wiley Interdisciplinary Reviews Climate Change* 14, no. 3 (2023): e824, 1–11

Collard, Rosemary-Claire. *Animal Traffic: Lively Capital in the Global Exotic Pet Trade*. Durham, NC: Duke University Press, 2020

Collard, Rosemary-Claire, and Jessica Dempsey. "Life for sale? The politics of lively commodities." *Environment and Planning A* 45, no. 11 (2013): 2682–2699

Collard, Rosemary-Claire, Jessica Dempsey, and Juanita Sundberg. "A manifesto for abundant futures." *Annals of the Association of American Geographers* 105, no. 2 (2015): 322–330

Committee on Climate Change. *Net Zero: The UK's Contribution to Stopping Global Warming*. London: Committee on Climate Change, 2019

Committee on Climate Change. *Land Use: Policies for a Net Zero UK*. London: Committee on Climate Change, 2020

Cooper, Melinda. *Life as Surplus: Biotechnology & Capitalism in the Neoliberal Era*. Seattle: University of Washington Press, 2008

Cooper, Melinda, and Catherine Waldby. *Clinical Labor: Tissue Donors and Research Subjects in the Global Bioeconomy*. Durham, NC: Duke University Press, 2014

Costello, Christine, W. Michael Griffin, Amy E. Landis, and H. Scott Matthews. "Impact of biofuel crop production on the formation of hypoxia in the Gulf of Mexico." *Environmental Science and Technology* 43, no. 20 (2009): 7985–7991

Cotula, Lorenzo, Nat Dyer, and Sonja Vermeulen. *Fuelling Exclusion? The Biofuels Boom and Poor People's Access to Land*. London: International Institute for Environment and Development, 2008

Cotula, Lorenzo, Sonja Vermeulen, Rebeca Leonard, and James Keeley. *Land Grab or Development Opportunity? Agricultural Investment and International Land Deals in Africa*. London: International Institute for Environment and Development, 2009

Croezen, Harry, Geert Bergsma, Matthijs Otten, and M.P.J. van Valkengoed. *Biofuels: Indirect Land Use Change and Climate Impact*. Delft: CE Delft, 2010

Cunningham, Kyle, John Barry, and Tamara Walkingstick. *Managing Loblolly Pine Stands ... from A to Z*. Little Rock: University of Arkansas Division of Agriculture Cooperative Extension Service, 2008

Daggett, Cara New. *The Birth of Energy: Fossil Fuels, Thermodynamics, & the Politics of Work*. Durham, NC: Duke University Press, 2019

Daggett, Cara New. "Energy and domination: Contesting the fossil myth of fuel expansion." *Environmental Politics* 30, no. 4 (2021): 644–662

Dauvergne, Peter, and Kate J. Neville. "Forests, food, and fuel in the tropics: The uneven social and ecological consequences of the emerging political economy of biofuels." *Journal of Peasant Studies* 37, no. 4 (2010): 631–660

Davies, Thom. "Slow violence and toxic geographies: 'Out of sight' to whom?" *Environment and Planning C: Politics and Space* 40, no. 2 (2022): 409–427

Demeritt, David. "Scientific forest conservation and the statistical picturing of nature's limits in the Progressive-era United States." *Environment and Planning D: Society and Space* 19 (2001): 431–459

Dempsey, Jessica. *Enterprising Nature: Economics, Markets, and Finance in Global Biodiversity Politics*. Oxford: Wiley-Blackwell, 2016

Doganova, Liliana. *Discounting the Future: The Ascendancy of a Political Technology*. Princeton, NJ: Princeton University Press, 2024

Dogwood Alliance. "Why is biomass bad? 5 reasons to stop importing bioenergy." 15 March 2022. https://dogwoodalliance.org/2022/03/5–reasons-why-the-world-must-stop-importing-bioenergy/ (accessed 19 March 2025)

Dogwood Alliance, National Resources Defense Council, and Southern Environmental Law Center. *European Imports of Wood Pellets for "Green Energy" Devastating US Forests*. 29 June 2017. https://www.nrdc.org/sites/default/files/european-imports-wood-pellets-greenenergy-devastating-us-forests.pdf (accessed 19 March 2025)

Dooley, Kate, Kristine Lund Christiansen, Jens Friis Lund, Wim Carton, and Alister Self. "Over-reliance on land for carbon dioxide removal in net-zero climate pledges." *Nature Communications* 15, no. 9118 (2024): 1–10.

Dooley, Kate, Peter Christoff, and Kimberly A. Nicholas. "Co-producing climate policy and negative emissions: Trade-offs for sustainable land-use." *Global Sustainability* 1, no. e3 (2018): 1–10

Doornbosch, Richard, and Ronald Steenblik. *Biofuels: Is the Cure Worse than the Disease?* Paris: Organisation for Economic Co-operation and Development, 2007

Drax Group plc. "5 things you never knew about forests." 6 October 2016. https://web.archive.org/web/20170410095804/https://www.drax.com/sustainability/5-things-you-never-knew-about-forests/ (accessed 21 March 2025)

Drax Group plc. "How wood pellets are revitalizing a community." 23 February 2017. https://www.drax.com/us/uncategorized/how-wood-pellets-are-revitalizing-a-community/ (accessed 17 March 2025)

Drax Group plc. "What is a working forest?" 1 June 2017. https://web.archive.org/web/20170930020916/https://www.drax.com/sustainability/what-is-a-working-forest/ (accessed 21 March 2025)

Drax Group plc. "Active management of forests increases growth and carbon storage." 28 July 2017. https://web.archive.org/web/20181021225054/https://www.drax.com/sustainability/active-management-forests-increases-growth-carbon-storage/ (accessed 21 March 2025)

Drax Group plc. "What can be made from captured carbon?" 11 September 2018. https://www.drax.com/technology/can-made-captured-carbon/ (accessed 24 March 2025)

Drax Group plc. "Carbon dioxide now being captured in first of its kind BECCS pilot." 7 February 2019. https://www.drax.com/press_release/world-first-co2-beccs-ccus/ (accessed 23 March 2025)

Drax Group plc. "New carbon capture technology could help industry and agricultural sector decarbonise." 17 June 2019. https://www.drax.com/press_release/new-carbon-capture-technology-help-industry-agricultural-sector-decarbonise/ (accessed 24 March 2025)

Drax Group plc. "Drax secures £500,000 for innovative fuel cell carbon capture study." 27 June 2019. https://www.drax.com/press_release/drax-secures-500000-innovative-fuel-cell-carbon-capture-study/ (accessed 24 March 2025)

Drax Group plc. "Drax's largest biomass shipment arrives at the UK's biggest biomass handling facility." 20 March 2020. https://www.drax.com/press_release/draxs-largest-biomass-shipment-arrives-at-the-uks-biggest-biomass-handling-facility/ (accessed 16 March 2025)

Drax Group plc. "Drax receives 100th biomass cargo from dedicated US export facility." 6 April 2020. https://www.drax.com/us/press_release/drax-receives-100th-biomass-cargo-from-dedicated-us-export-facility/ (accessed 16 March 2025)

Drax Group plc. "Jobs, skills, zero emissions: The economic need for carbon capture by Drax." 18 November 2020. https://www.drax.com/carbon-capture/jobs-skills-zero-emissions-the-economic-need-for-carbon-capture-by-drax/ (accessed 24 March 2025)

Drax Group plc. *Committed to the World's Energy Transition: Annual Report and Accounts 2023*. Selby: Drax Group plc, 2024. https://www.drax.com/wp-content/uploads/2024/03/Drax_AR23_Interactive.pdf#page=19 (accessed 17 March 2025)

Dukes, Jeffrey S. "Burning buried sunshine: Human consumption of ancient solar energy." *Climatic Change* 61, nos. 1–2 (2003): 33–41

Dunlap, Alexander. *Renewing Destruction: Wind Energy Development, Conflict and Resistance in a Latin American Context*. London: Rowman and Littlefield International, 2019

Dunlap, Alexander, and Martín Correa Arce. "'Murderous energy' in Oaxaca, Mexico: Wind factories, territorial struggle, and social warfare." *Journal of Peasant Studies* 49, no. 2 (2022): 455–480

Eaglin, Jennifer. "The demise of the Brazilian ethanol program: Environmental and economic shocks, 1985–1990." *Environmental History* 24, no. 1 (2019): 104–129

Eaglin, Jennifer. *Sweet Fuel: A Political and Environmental History of Brazilian Ethanol*. Oxford: Oxford University Press, 2022

Edelman, Marc, Carlos Oya, and Saturnino M. Borras Jr. "Global land grabs: Historical processes, theoretical and methodological implications and current trajectories." *Third World Quarterly* 34, no. 9 (2013): 1517–1531

Eden, Sally. *Environmental Publics*. Abingdon: Routledge, 2017

Edwards, Robert, Declan Mulligan, and Luisa Marelli. *Indirect Land Use Change from Increased Biofuel Demand: Comparison of Models and Results for Marginal Biofuels Production from Different Feedstocks*. Ispra: European Commission Joint Research Centre, 2010

Edwards, Robert, Szabolcs Szekeres, Frederik Neuwahl, and Vincent Mahieu. *Biofuels in the European Context: Facts and Uncertainties*. Ispra: European Commission Joint Research Centre, 2008

Ekers, Michael, and Scott Prudham. "The metabolism of socioecological fixes: Capital switching, spatial fixes, and the production of nature." *Annals of the American Association of Geographers* 107, no. 6 (2017): 1370–1388

Energy Institute. *Statistical Review of World Energy 2024*. London: Energy Institute, 2024. https://www.energyinst.org/statistical-review (accessed 26 September 2024)

Energy Technologies Institute. *Bioenergy: Delivering Greenhouse Gas Emissions Savings through UK Bioenergy Value Chains*. Loughborough: Energy Technologies Institute, 2015

Environmental Integrity Project. *Dirty Deception: How the Wood Biomass Industry Skirts the Clean Air Act*. 26 April 2018. https://environmentalintegrity.org/wp-content/uploads/2017/02/Biomass-Report.pdf (accessed 21 March 2025)

Ernwein, Marion. *Les Natures de la Ville Néolibérale: Une écologie politique du végétal urbain*. Grenoble: UGA Editions, 2019

Ernwein, Marion. "From undead commodities to lively labourers: (Re)valuing vegetal life, reclaiming the power to design-with plants." In *The Botanical City*, edited by Matthew Gandy and Sandra Jasper, 237–242. Berlin: Jovis Verlag, 2020

Ernwein, Marion. "Bringing urban parks to life: The more-than-human politics of urban ecological work." *Annals of the American Association of Geographers* 111, no. 2 (2021): 559–576

Ernwein, Marion, Franklin Ginn, and James Palmer, eds, *The Work That Plants Do: Life, Labour and the Future of Vegetal Economies*. Berlin: Transcript Verlag, 2021

Ernwein, Marion, and James Palmer. "Making the mos(s)t of nature: Cleantech, smart nature-based solutions, and the 'rendering investable' of urban moss." *Environment and Planning E: Nature and Space* 8, no. 1 (2025): 306–324

Ernwein, Marion, James Palmer, and Franklin Ginn. "Introduction: The work that plants do." In *The Work That Plants Do: Life, Labour and the Future of Vegetal Economies*, edited by Marion Ernwein, Franklin Ginn, and James Palmer, 15–31. Berlin: Transcript Verlag, 2021

Escobar, José C., Electo S. Lora, Osvaldo J. Venturini *et al.* "Biofuels: Environment, technology and food security." *Renewable and Sustainable Energy Reviews* 13, nos. 6–7 (2009): 1275–1287

European Commission. *European Transport Policy for 2010: Time to Decide, COM (2001) 730 Final*. Brussels: European Commission, 2001

European Commission. "Directive 2003/30/EC of the European Parliament and of the Council of 8 May 2003 on the promotion of the use of biofuels or other renewable fuels for transport." *Official Journal of the European Union* L123 (2003): 42–46. Brussels: European Commission

European Commission. *Annex to the Impact Assessment Document Accompanying the Package of Implementation Measures for the EU's Objectives on Climate Change and Renewable Energy for 2020, SEC (2008) 85 Vol. II*. Brussels: European Commission, 2008

European Commission. "Directive 2009/28/EC of the European Parliament and of the Council of 23 April 2009 on the promotion of the use of energy from renewable sources and amending and subsequently repealing Directives 2001/77/EC and 2003/30/EC." *Official Journal of the European Union* L140 (2009): 16–61. Brussels: European Commission

European Commission. "Directive (EU) 2018/2001 of the European Parliament and of the Council of 11 December 2018 on the promotion and use of energy from renewable resources (recast)." *Official Journal of the European Union* L328 (2018): 82–209

European Commission. *A Sustainable Bioeconomy for Europe: Strengthening the Connection between Economy, Society, and the Environment*. Brussels: European Commission Directorate-General for Research and Innovation, 2018

Fair, Hannah, and Matthew McMullen. "Toward a theory of nonhuman species-being." *Environmental Humanities* 15, no. 2 (2023):195–214

Fajardy, Mathilde, and Niall MacDowell. "Can BECCS deliver sustainable and resource efficient negative emissions?" *Energy and Environmental Science* 10 (2017): 1389–1426

FAO. *Land Statistics 2001–2022: Global, regional and country trends*. Rome: Food and Agriculture Organization of the United Nations, 2024

Fargione, Joseph, Jason Hill, David Tilman *et al*. "Land clearing and the biofuel carbon debt." *Science* 319 (2008): 1235–1238

Federici, Silvia. *Wages against Housework*. Bristol: Falling Wall Press, 1975

Fergusson, Malcolm, Ausilio Bauen, Nick Eyre *et al*. *Expert Paper on the Global Impacts of Road Transport Biofuels: A Contribution to the Government's Analysis*. London: The Institute for European Environmental Policy, 2004

Fish, Kenneth. *Living Factories: Biotechnology and the Unique Nature of Capitalism*. Montreal: McGill-Queen's University Press, 2013

Fitzgerald, Deborah. *Every Farm a Factory: The Industrial Ideal in American Agriculture*. New Haven, CT: Yale University Press, 2010

Fitzsimmons, Margaret. "The matter of nature." *Antipode* 21, no. 2 (1989): 106–120

Forest2Market. *Historical Perspective on the Relationship between Demand and Forest Productivity in the US South: At a Glance*. July 2017. https://www.forest2market.com/hubfs/2016_Website/Documents/20170726_Forest2Market_Historical_Perspective_US_South_At_a_Glance.pdf (accessed 17 March 2025)

Forestry Commission England. *Sustainable Forest Management*. November 2018. https://www.theccc.org.uk/wp-content/uploads/2018/11/Annex-1-Sustainable-Forest-Management.pdf (accessed 23 March 2025)

Forzieri, Giovanni, Vasilis Dakos, Nate G. McDowell *et al*. "Emerging signals of declining forest resilience under climate change." *Nature* 608 (2022): 534–539

Foster, John Bellamy, and Brett Clark. "The paradox of wealth: Capitalism and ecological destruction." *Monthly Review* 61, no. 6 (2009): 1–18
Foucault, Michel. "The confession of the flesh." In *Power/Knowledge: Selected Interviews and Other Writings, 1972–1977*, edited by Colin Gordon, 194–228. New York: Pantheon, 1980
Foucault, Michel. *The Order of Things: An Archaeology of the Human Sciences*. Abingdon: Routledge, 2002[1970]
Fougeres, Sophie. "First solely-biofuel jet flight raises clean travel hopes." *Aerospace and Defense News*, 8 November 2012. https://web.archive.org/web/20121114032038/http:/www.asdnews.com/news-46032/First_solely-biofuel_jet_flight_raises_clean_travel_hopes.htm (accessed 31 January 2025)
Foulsham, George. "UCLA researchers turn carbon dioxide into sustainable concrete." *UCLA Newsroom*, 14 March 2016. https://newsroom.ucla.edu/releases/ucla-researchers-turn-carbon-dioxide-into-sustainable-concrete (accessed 24 March 2025)
Fressoz, Jean-Baptise. *More and More and More: An All-Consuming History of Energy*. London: Allen Lane, 2024
Gabrys, Jennifer. "Sink: The dirt of systems." *Environment and Planning D: Society and Space* 27, no. 4 (2009): 666–681
Geden, Oliver. "The Paris Agreement and the inherent inconsistency of climate policymaking." *Wiley Interdisciplinary Reviews Climate Change* 7 (2016): 790–797
Geden, Oliver, Vivian Scott, and James Palmer. "Integrating carbon dioxide removal into EU climate policy: Prospects for a paradigm shift." *Wiley Interdisciplinary Reviews Climate Change* 9, no. e521 (2018): 1–10
Geels, Frank, Benjamin K. Sovacool, Tim Schwanen, and Steve Sorrell. "Sociotechnical transitions for deep decarbonisation." *Science* 357, no. 6357 (2017): 1242–1244
Georgiadis, Philip, and Jim Pickard. "UK forecasts small role for green airline fuel by 2040 despite Reeves' claims." *The Financial Times*, 30 January 2025. https://www.ft.com/content/78c3705d-d484-48ce-adfb-ffc827fc0646 (accessed 8 March 2025)
Giampietro, Mario. "On the circular bioeconomy and decoupling: Implications for sustainable growth." *Ecological Economics* 162 (2019): 143–156
Giampietro, Mario, and Kozo Mayumi. *The Biofuel Delusion: The Fallacy of Large-Scale Agro-biofuel Production*. Abingdon: Routledge, 2009
Gibbs, Holly K., and J. Meghan Salmon. "Mapping the world's degraded lands." *Applied Geography* 57 (2015): 12–21
Gill, Rosalind, and Andy Pratt. "In the social factory? Immaterial labour, precariousness and cultural work." *Theory, Culture and Society* 25, nos. 7–8 (2008): 1–30
Ginn, Franklin. "Extension, subversion, containment: Eco-nationalism and (post) colonial nature in Aotearoa New Zealand." *Transactions of the Institute of British Geographers* 33, no. 3 (2008): 335–353
Ginn, Franklin, and Katy Connor. "Vegetal hydropoetics: An arts-based practice for plant studies." *Cultural Geographies* 30, no. 3 (2023): 493–497
Ginn, Franklin, and Daanish Mustafa. "Planting soft Pakistan." In *The Work That Plants Do: Life, Labour and the Future of Vegetal Economies*, edited by Marion Ernwein, Franklin Ginn, and James Palmer, 71–83. Berlin: Transcript Verlag, 2021
Glasgow Declaration on Sustainable Bioenergy. *Sustainable Bioenergy at the Heart of Global Net Zero*. November 2021. https://sustainablebioenergy.org/declaration/ (accessed 4 February 2025)

Goldstein, Jesse. "*Terra economica*: Waste and the production of enclosed nature." *Antipode* 45, no. 2 (2013): 357–375

Goldstein, Jesse. "Net zero and settler futurity." *New Formations* 112 (2024): 31–50

Goldstein, Jesse, and Elizabeth Johnson. "Biomimicry: New natures, new enclosures." *Theory, Culture & Society* 32, no. 1 (2015): 61–81

Goodman, David, Bernardo Sorj, and John Wilkinson. *From Farming to Biotechnology: A Theory of Agro-Industrial Development*. London: Basil Blackwell, 1987

Graeber, David. *Bullshit Jobs: A Theory*. London: Allen Lane, 2018

Hansen, M.C., P.V. Potapov, R. Moore *et al.* "High-resolution global maps of 21st-century forest cover change." *Science* 342, no. 6160 (2013): 850–853

Haraway, Donna. *When Species Meet*. Minneapolis: University of Minnesota Press, 2007

Harlan, Tyler, and Jennifer Baka. "Stacked energyscapes: Conceptualizing fossil fuel and renewable energy entanglements in low-carbon transitions." *Energy Research & Social Science* 115, no. 103648 (2024): 1–10

Harp, Stephen. *A World History of Rubber: Empire, Industry, and the Everyday*. Oxford: Wiley-Blackwell, 2016

Hayes, Denis. "Plant power: Biological sources of energy." *The Ecologist: Journal of the Post Industrial Age* 7, no. 9 (1977): 349–356

Head, Lesley, Jennifer Atchison, Catherine Phillips, and Kathleen Buckingham. "Vegetal politics: Belonging, practices and places." *Social & Cultural Geography* 15, no. 8 (2014): 861–870

Helm, Dieter. *Natural Capital: Valuing the Planet*. New Haven, CT: Yale University Press, 2015

Henry, Robert. *Plant Resources for Food, Fuel and Conservation*. London: Earthscan, 2010

Hickel, Jason. "The anti-colonial politics of degrowth." *Political Geography* 88, no. 102404 (2021): 1–3

Hickman, Leo. "How BECCS became climate change's 'saviour' technology." *Carbon Brief*, 13 April 2016. https://www.carbonbrief.org/beccs-the-story-of-climate-changes-saviour-technology/ (accessed 23 March 2025)

Hitch, Gregory, and Marcus Grignon. "A forest of energy: Settler colonialism, knowledge production, and sugar maple kinship in the Menominee community." *American Quarterly* 75, no. 2 (2023): 251–277

Hitchner, Sarah, John Schelhas, and J. Peter Brosius. *Forests as Fuel: Energy, Landscape, Climate, and Race in the U.S. South*. Lanham, MD: Lexington Books, 2022

Hochschild, Arlie Lee. *The Managed Heart: Commercialization of Human Feeling*. Berkeley: University of California Press, 2012

Hodges, Donald G., A.J. Hartsell, C. Brandeis *et al.* "Recession effects on the forests and forest product industries of the South." *Forest Products Journal* 61, no. 8 (2011): 614–624

Howarth, Robert Warren, and Stefan Bringezu, eds, *Biofuels: Environmental Consequences and Interactions with Changing Land Use*. Proceedings of the Scientific Committee on Problems of the Environment (SCOPE) International Biofuels Project Rapid Assessment, 22–25 September 2008, Gummersbach, Germany. Ithaca, NY: Cornell University Library eCommons, 2009

Howse, Robert, Petrus van Bork, and Charlotte Hebebrand. *WTO Disciplines and Biofuels: Opportunities and Constraints in the Creation of a Global Marketplace.*

IPC Discussion Paper. Washington, DC: International Policy Council and Renewable Energy and International Law (REIL), 2006

Hribal, Jason. "'Animals are part of the working class': A challenge to labor history." *Labor History* 44, no. 4 (2003): 435–453

Huber, Matt. *Lifeblood: Oil, Freedom, and the Forces of Capital.* Minneapolis: University of Minnesota Press, 2013

Huber, Matt. "Energized antagonisms: Thinking beyond 'energy culture'." In *Energy Culture: Art and Theory on Oil and Beyond*, edited by Imre Szeman and Jeff Diamanti, 233–245. Morgantown: West Virginia University Press, 2019

Huber, Matt, and James McCarthy. "Beyond the subterranean energy regime? Fuel, land use and the production of space." *Transactions of the Institute of British Geographers* 42 no. 4 (2017): 655–668

Hunsberger, Carol, and Stefano Ponte. "'Sustainable' biofuels in the global South." *Geoforum* 54 (2014): 243–247

Intergovernmental Panel on Climate Change. *Climate Change 2014 Synthesis Report.* Geneva: World Meteorological Organization, 2015

Intergovernmental Panel on Climate Change. *Global Warming of 1.5°C: An IPCC Special Report. Summary for Policymakers.* Geneva: World Meteorological Organization, 2018

Intergovernmental Panel on Climate Change. *Climate Change 2022: Mitigation of Climate Change. Summary for Policymakers.* Geneva: World Meteorological Organization, 2022

International Air Transport Association. "Net-zero carbon emissions by 2050." 4 October 2021. https://www.iata.org/en/pressroom/pressroom-archive/2021-releases/2021-10-04-03/ (accessed 15 March 2025)

International Air Transport Association. *Net Zero 2050: Sustainable Aviation Fuels Fact Sheet.* May. Montreal: International Air Transport Association, 2024

International Energy Agency. *Biofuels for Transport: An International Perspective.* Paris: International Energy Agency, 2004

International Energy Agency. "Modern bioenergy leads the growth of all renewables to 2023, according to latest IEA market forecast." 8 October 2018. https://www.iea.org/news/modern-bioenergy-leads-the-growth-of-all-renewables-to-2023-according-to-latest-iea-market-forecast (accessed 2 February 2025)

International Energy Agency. *Renewables 2023: Analysis and Forecast to 2028.* Paris: International Energy Agency Publications, 2024

International Monetary Fund. *The Impact of Higher Oil Prices on the Global Economy.* 8 December 2000. https://www.imf.org/external/pubs/ft/oil/2000/oilrep.PDF (accessed 15 September 2024)

International Renewable Energy Agency. *Global Renewables Outlook: Energy Transformation 2050.* Abu Dhabi: International Renewable Energy Agency, 2020

Irigaray, Luce. "What the vegetal world says to us." In *The Language of Plants: Science, Philosophy, Literature*, edited by Monica Gagliano, John C. Ryan, and Patrícia Vieira, 126–135. Minneapolis: University of Minnesota Press, 2019

Iuel-Stissing, Jens, Peter Karnøe, and Susse Georg. "Biomass qualification and the dynamics of market framing." In *Market Studies: Mapping, Theorizing and Impacting Market Action*, edited by Susi Geiger, Katy Mason, Neil Pollock *et al.*, 176–193. Cambridge: Cambridge University Press, 2024

Jevons, William Stanley. *The Coal Question: An Inquiry Concerning the Progress of the Nation, and the Probable Exhaustion of Our Coal Mines.* London: Macmillan and Co., 1865

Jonsson, Frederick Albritton. "Abundance and scarcity in geological time, 1784–1844." In *Nature, Action and the Future: Political Thought and the Environment*, edited by Katrina Forrester and Sophie Smith, 70–93. Cambridge: Cambridge University Press, 2018

Juskus, Ryan. "Sacrifice zones: A genealogy and analysis of an environmental justice concept." *Environmental Humanities* 15, no. 1 (2023): 3–24

Kallis, Giorgos. *Limits: Why Malthus Was Wrong and Why Environmentalists Should Care*. Stanford, CA: Stanford University Press, 2019

Kallis, Giorgos, and Erik Swyngedouw. "Do bees produce value? A conversation between an ecological economist and a Marxist geographer." *Capitalism, Nature, Socialism* 29, no. 3 (2018): 36–50

Kama, Kärg. "Resource-making controversies: Knowledge, anticipatory politics and economization of unconventional fossil fuels." *Progress in Human Geography* 44, no. 2 (2020): 333–356

Kareiva, Peter, Heather Tallis, Taylor H. Ricketts *et al.*, eds, *Natural Capital: Theory and Practice of Mapping Ecosystem Services*. Oxford: Oxford University Press, 2011

Kasdogan, Duygu. *Potentiating Algae, Modernizing Bioeconomies: Algal Biofuels, Bioenergy Economies, and Built Ecologies in the United States and Turkey*. PhD diss., York University Toronto, 2017

Keeney, Dennis. "Ethanol USA." *Environmental Science & Technology* 43, no. 1 (2009): 8–11

Kimmerer, Robin Wall. *Braiding Sweetgrass: Indigenous Wisdom, Scientific Knowledge, and the Teachings of Plants*. London: Penguin Books, 2013

Kingdon, John. *Agendas, Alternatives, and Public Policies*. New York: Longman, 2003

Kirsch, Scott, and Don Mitchell. "The nature of things: Dead labor, nonhuman actors, and the persistence of Marxism." *Antipode* 36, no. 4 (2004): 687–705

Kloppenburg, Jack. *First the Seed: The Political Economy of Plant Biotechnology*. Madison: University of Wisconsin Press, 2004

Knuth, Sarah. "Green devaluation: Disruption, divestment, and decommodification for a green economy." *Capitalism Nature Socialism* 28, no. 1 (2017): 98–117

Koester, Stefan, and Sam Davis. "Siting of wood pellet production facilities in environmental justice communities in the southeastern United States." *Environmental Justice* 11, no. 2 (2018): 64–70

Koh, Lian Pin, Jukka Miettinen, Soo Chin Liew, and Jaboury Ghazoul. "Remotely sensed evidence of tropical peatland conversion to oil palm." *Proceedings of the National Academy of Sciences* 108, no. 12 (2011): 5127–5132

Kohn, Eduardo. *How Forests Think: Toward an Anthropology beyond the Human*. Berkeley: University of California Press, 2013

Kruger, Tim, Oliver Geden, and Steve Rayner. "Abandon hype in climate models." *The Guardian*, 26 April 2016. https://www.theguardian.com/science/political-science/2016/apr/26/abandon-hype-in-climate-models?alm_mvr=0 (accessed 24 March 2025)

Krzywoszynska, Anna. "Nonhuman labor and the making of resources: Making soils a resource through microbial labor." *Environmental Humanities* 12, no. 1 (2020): 227–249

Kuchler, Magdalena. "Sweet dreams (are made of cellulose): Sociotechnical imaginaries of second-generation bioenergy in the global debate." *Ecological Economics* 107 (2014): 431–437

Kugelman, Michael, and Susan L. Levenstein, eds, *The Global Farms Race: Land Grabs, Agricultural Investment, and the Scramble for Food Security*. Washington, DC: Island Press, 2012

Kumar, Deepak, Stephen P. Long, and Vijay Singh. "Biorefinery for combined production of jet fuel and ethanol from lipid-producing sugarcane: A techno-economic evaluation." *Global Change Biology Bioenergy* 10, no. 2 (2018): 92–107

Kutas, Géraldine, Carina Lindberg, and Ronald Steenblik. *Biofuels—At What Cost? Government Support for Ethanol and Biodiesel in the European Union*. Geneva: Global Subsidies Initiative, 2007

Lamb, William F., Giulio Mattoli, Sebastian Levi *et al.* "Discourses of climate delay." *Global Sustainability* 3 (2020): 1–5

Lawrence, Anna. "Listening to plants: Conversations between critical plant studies and vegetal geography." *Progress in Human Geography* 46, no. 2 (2022): 629–651

Lennon, Myles. "Decolonizing energy: Black Lives Matter and technoscientific expertise amid solar transitions." *Energy Research & Social Science* 30 (2017): 18–27

Lennon, Myles. "Postcarbon amnesia: Toward a recognition of racial grief in renewable energy futures." *Science, Technology, & Human Values* 45, no. 5 (2020): 934–962

Lenton, Timothy M., and Bruno Latour. "Gaia 2.0." *Science* 361, no. 6407 (2018): 1066–1068

Levenda, Anthony, Ingrid Behrsin, and F. Disano. "Renewable energy for whom? A global systematic review of the environmental justice implications of renewable energy technologies." *Energy Research & Social Science* 71, no. 101837 (2021): 1–13

Levidow, Les. "EU criteria for sustainable biofuels: Accounting for carbon, depoliticising plunder." *Geoforum* 44 (2013): 211–233

Lewis, Simon L., Charlotte E. Wheeler, Edward T.A. Mitchard, and Alexander Koch. "Regenerate natural forests to store carbon." *Nature* 568 (2019): 25–28

Li, Tania Murray. "To make live or let die? Rural dispossession and the protection of surplus populations." *Antipode* 41, no. S1 (2010): 66–93

Li, Tania Murray. "What is land? Assembling a resource for global investment." *Transactions of the Institute of British Geographers* 39 (2014): 589–602

Lohmann, Larry. "Bioenergy, thermodynamics and inequalities." In *Bioeconomy and Global Inequalities: Socio-Ecological Perspectives on Biomass Sourcing and Production*, edited by Maria Backhouse, Rosa Lehmann, Kristina Lorenzen *et al.*, 85–103. Basingstoke: Palgrave Macmillan, 2021

Lustgarten, Abrahm. "Palm oil was supposed to help save the planet. Instead it unleashed a catastrophe." *New York Times Magazine*, 20 November 2018. https://www.nytimes.com/2018/11/20/magazine/palm-oil-borneo-climate-catastrophe.html (accessed 1 October 2024)

Lyons, Kristina M. *Vital Decomposition: Soil Practitioners and Life Politics*. Durham, NC: Duke University Press, 2020

Malm, Andreas. *Fossil Capital: The Rise of Steam Power and the Roots of Global Warming*. London: Verso, 2016

Malm, Andreas, and Wim Carton. *Overshoot: How the World Surrendered to Climate Breakdown*. London: Verso, 2024

Mancuso, Stefano. *The Revolutionary Genius of Plants: A New Understanding of Plant Intelligence and Behavior*. New York: Atria, 2018

Manuel, Jeffrey T. "Lessons from a forgotten fuel: Assessing the long history of alcohol fuel advocacy and use in the United States." *History and Technology* 37, no. 4 (2021): 411–428

Marder, Michael. *Plant-Thinking: A Philosophy of Vegetal Life*. New York: Columbia University Press, 2013

Marder, Michael. "The sense of seeds, or seminal events." *Environmental Philosophy* 12, no. 1 (2015): 87–98

Marder, Michael. "To hear plants speak." In *The Language of Plants: Science, Philosophy, Literature*, edited by Monica Gagliano, John C. Ryan, and Patrícia Vieira, 103–125. Minneapolis: University of Minnesota Press, 2019

Marx, Karl. *Capital: A Critique of Political Economy*, vol. I. London: Penguin, 1976[1867]

Matondi, Prosper B., Kjell Havnevik, and Atakilte Beyene, eds, *Biofuels, Land Grabbing and Food Security in Africa*. London: Zed Books, 2011

Matthews, Robert, Geoff Hogan, and Ewan Mackie. *Carbon Impacts of Biomass Consumed in the EU: Supplementary Analysis and Interpretation for the European Climate Foundation*. Forest Research, May 2018. https://europeanclimate.org/wp-content/uploads/2018/05/CIB-Summary-report-for-ECF-v10.5-May-20181.pdf (accessed 17 March 2025)

Mayr, Stefan, Birgit Hollaus, and Verena Madner. "Palm oil, the RED II and WTO law: EU sustainable biofuel policy tangled up in green?" *Review of European, Comparative and International Environmental Law* 30, no. 2 (2021): 233–248

McCarthy, James. "A socioecological fix to capitalist crisis and climate change? The possibilities and limits of renewable energy." *Environment and Planning A: Economy and Space* 47, no. 12 (2015): 2485–2502

McCarthy, Tom. *Auto Mania: Cars, Consumers, and the Environment*. New Haven, CT: Yale University Press, 2007

McDonagh, John. "Rural geography III: Do we really have a choice? The bioeconomy and future rural pathways." *Progress in Human Geography* 39, no. 5 (2015): 658–665

McKeand, Steven E., Eric J. Jokela, Dudley A. Huber *et al.* "Performance of improved genotypes of loblolly pine across different soils, climates, and silvicultural inputs." *Forest Ecology and Management* 227, nos. 1–2 (2006): 178–184

McMichael, Philip. "A food regime analysis of the 'world food crisis'." *Agriculture and Human Values* 26 (2009): 281–295

Mission Possible Partnership. *Making Net-Zero Aviation Possible: An Industry-Back, 1.5°C-Aligned Transition Strategy*, July 2022. https://3stepsolutions.s3-accelerate.amazonaws.com/assets/custom/010856/downloads/Making-Net-Zero-Aviation-possible.pdf (accessed 15 March 2025)

Mitchell, Donald. *A Note on Rising Food Prices*. Policy Research Working Paper 4682. Washington, DC: World Bank Development Prospects Group, 2008

Mitchell, Timothy. *Rule of Experts: Egypt, Techno-Politics, Modernity*. Berkeley: University of California Press, 2002

Mitchell, Timothy. "Carbon democracy." *Economy and Society* 38, no. 3 (2009): 399–432

Mitchell, Timothy. *Carbon Democracy: Political Power in the Age of Oil*. London: Verso, 2011

Moffat, Anne Simon. "Plants as chemical factories." *Science* 268, no. 5211 (1995): 659

Mol, Arthur P.J. "Boundless biofuels? Between environmental sustainability and vulnerability." *Sociologia Ruralis* 47, no. 4 (2007): 297–315

Monbiot, George. *Heat: How We Can Stop the Planet Burning*. London: Penguin, 2007

Moomaw, William R., Susan A. Masino, and Edward K. Faison. "Intact forests in the United States: Proforestation mitigates climate change and serves the greatest good." *Frontiers in Forests and Global Change* 2, no. 27 (2019): 1–10

Moore, Jason W. "Transcending the metabolic rift: A theory of crises in the capitalist world-ecology." *Journal of Peasant Studies* 38, no. 1 (2011): 1–46

Moore, Jason W. *Capitalism in the Web of Life: Ecology and the Accumulation of Capital*. London: Verso, 2015

Morton, Oliver. *Eating the Sun: How Plants Power the Planet*. London: Fourth Estate, 2007

Myers, Natasha. "Photosynthesis." *Cultural Anthropology: Theorizing the Contemporary*, 21 January 2016. https://culanth.org/fieldsights/photosynthesis (accessed 30 January 2025)

Myers, Natasha. "Photosynthetic mattering: Rooting into the Planthroposcene." In *Moving Plants*, edited by Line Marie Thorsen, 123–129. Næstved, Denmark: Rønnebæksholm, 2017

Myers, Natasha. "From edenic apocalypse to gardens against Eden: Plants and people in and after the Anthropocene." In *Infrastructure, Environment, and Life in the Anthropocene*, edited by Kregg Hetheringon, 115–148. Durham, NC: Duke University Press, 2019

Nally, David. "Governing precarious lives: Land grabs, geopolitics, and 'food security'." *The Geographical Journal* 181, no. 4 (2015): 340–349

Nansen, Christian. "Unraveling the 'radiometric signals' from green leaves." In *The Language of Plants: Science, Philosophy, Literature*, edited by Monica Gagliano, John C. Ryan, and Patrícia Vieira, 62–83. Minneapolis: University of Minnesota Press, 2019

National Renewable Energy Laboratory. *Sustainable Aviation Fuel (SAF) State-of-Industry Report: State of SAF Production Process*, July 2024. https://docs.nrel.gov/docs/fy24osti/87802.pdf (accessed 12 December 2025)

Nealon, Jeffrey T. *Plant Theory: Biopower & Vegetable Life*. Stanford, CA: Stanford University Press, 2016

Neville, Kate. *Fueling Resistance: The Contentious Political Economy of Biofuels and Fracking*. Oxford: Oxford University Press, 2021

Neville, Kate. *Going to Seed: Essays in Idleness, Nature & Sustainable Work*. Lubbock: Texas Tech University Press, 2024

Nixon, Rob. *Slow Violence and the Environmentalism of the Poor*. Cambridge, MA: Harvard University Press, 2011

Obersteiner, Michael, Christian Azar, Kenneth Möllersten *et al. Biomass Energy, Carbon Removal and Permanent Sequestration—A "Real Option" for Managing Climate Risk*. International Institute for Applied Systems Analysis, 20 January 2002. https://pure.iiasa.ac.at/id/eprint/6743/1/IR-02-042.pdf (accessed 23 March 2025)

O'Connor, James. "On the two contradictions of capitalism." *Capitalism Nature Socialism* 2, no. 3 (1991): 107–109

Office for National Statistics. "A burning issue: Biomass is the biggest source of renewable energy consumed in the UK." https://www.ons.gov.uk/economy/environmentalaccounts/articles/aburningissuebiomassisthebiggestsourceofrenewableenergyconsumedintheuk/2019-08-30 (accessed 23 March 2025)

Oliveira, Gustavo de L.T., Ben McKay, and Christina Plank. "How biofuel policies backfire: Misguided goals, inefficient mechanisms, and political-ecological blind spots." *Energy Policy* 108 (2017): 765–775

Oliver, Chadwick Dearing, Nedal T. Nassar, Bruce R. Lippke, and James B. McCarter. "Carbon, fossil fuel, and biodiversity mitigation with wood and forests." *Journal of Sustainable Forestry* 33, no. 3 (2014): 248–275

Oomen, Jeroen, Jess Hoffman, and Maarten A. Hajer. "Techniques of futuring: On how imagined futures become socially performative." *European Journal of Social Theory* 25, no. 2 (2022): 252–270

Oswalt, Sonja N., and W. Brad Smith. *U.S. Forest Resource Facts and Historical Trends*. Report FS-1036, August. Washington, DC: United States Department of Agriculture Forest Service, 2014

Oswalt, Sonja N., W. Brad Smith, Patrick D. Miles, and Scott A. Pugh. *Forest Resources of the United States, 2017: A Technical Document Supporting the Forest Service 2020 RPA Assessment*. Forest Service Gen Tech Report WO-97, March. Washington, DC: United States Department of Agriculture Forest Service, 2019

Owens, Susan. "'Engaging the public': Information and deliberation in environmental policy." *Environment and Planning A* 32, no. 7 (2000): 1141–1148

Owens, Susan. *Knowledge, Policy, and Expertise: The UK Royal Commission on Environmental Pollution, 1970–2011*. Oxford: Oxford University Press, 2015.

Padella, Monica, Adrian O'Connell, and Matteo Prussi. "What is still limiting the deployment of cellulosic ethanol? An analysis of the current status of the sector." *Applied Sciences* 9, no. 21 (2019): 4523

Palmer, James. "Stopping the unstoppable? A discursive-institutionalist analysis of renewable transport fuel policy." *Environment and Planning C: Government and Policy* 28, no. 6 (2010): 992–1010

Palmer, James. *Science and Politics in European Energy and Environmental Policy: The Wicked Problem of Biofuels and Indirect Land-Use Change (ILUC)*. PhD diss., University of Cambridge, 2013

Palmer, James. "Forest fuels: Vegetal labour and the reinvention of working forests as carbon conveyors in the US South." In *The Work That Plants Do: Life, Labour and the Future of Vegetal Economies*, edited by Marion Ernwein, Franklin Ginn, and James Palmer, 163–179. Berlin: Transcript Verlag, 2021

Palmer, James. "Putting forests to work? Enrolling vegetal labor in the socioecological fix of bioenergy resource making." *Annals of the American Association of Geographers* 141, no. 1 (2021): 141–156

Palmer, James. "Rethinking plant power: Bioenergy, vegetal labor, and postproductivist energy cultures." *Environmental Humanities* 16, no. 3 (2024): 661–679

Palmer, James, and Wim Carton. "Carbon removal as carbon revival? Bioenergy, negative emissions, and the politics of alternative energy futures." *Frontiers in Climate* 3 (2021): 1–7

Paprocki, Kasia, and James McCarthy. "The agrarian question of climate change." *Progress in Human Geography* 48, no. 6 (2024): 691–705

Parajuli, Saroj, Baskaran Kannan, Ratna Karan *et al.* "Towards oilcane: Engineering hyperaccumulation of triacylglycerol into sugarcane stems." *GCB-Bioenergy* 12, no. 7 (2020): 476–490

Park, Kiyoul, Truyen Quach, Teresa J. Clark *et al.* "Development of vegetative oil sorghum: From lab-to-field." *Plant Biotechnology Journal* 23, no. 2 (2025): 660–673

Paxson, Heather. "The naturalization of nature as working." *Cultural Anthropology: Theorizing the Contemporary*, 26 July 2018. https://culanth.org/fieldsights/the-naturalization-of-nature-as-working (accessed 21 March 2025)

Peltola, Taru, and Johanna Tuomisaari. "Making a difference: Forest biodiversity, affective capacities, and the micro-politics of expert fieldwork." *Geoforum* 64 (2015): 1–11

Perec, Georges. *A Man Asleep*. London: Vintage, 2011[1967]

Perkins, Harold. "Ecologies of actor-networks and (non)social labor within the urban political economies of nature." *Geoforum* 38 (2007): 1152–1162

Pollan, Michael. "The intelligent plant." *New Yorker*, 23 December 2013

Ponte, Stefano. "'Roundtabling' sustainability: Lessons from the biofuel industry." *Geoforum* 54 (2014): 261–271

Ponting, Clive. *A New Green History of the World: The Environment and the Collapse of Great Civilisations*. London: Vintage, 2007

Porcher, Jocelyne, and Jean Estebanez, eds, *Animal Labor: A New Perspective on Human-Animal Relations*. Berlin: Transcript Verlag, 2019

Porcher, Jocelyne, and Tiphaine Schmitt. "Dairy cows: Workers in the shadows?" *Society & Animals* 20, no. 1 (2012): 39–60

Postone, Moishe. *Time, Labor, and Social Domination: A Reinterpretation of Marx's Critical Theory*. Cambridge: Cambridge University Press, 1993

Pratley, Nils. "Drax is the subsidy show that goes on and on." *The Guardian*, 10 February 2025. https://www.theguardian.com/business/nils-pratley-on-finance/2025/feb/10/drax-is-the-subsidy-show-that-goes-on-and-on (accessed 17 March 2025)

Proskurina, Svetlana, Martin Junginger, Jussi Heinimö *et al.* "Global biomass trade for energy—Part 2: Production and trade streams of wood pellets, liquid biofuels, charcoal, industrial roundwood and emerging energy biomass." *Biofuels, Bioproducts and Biorefining* 13 (2019): 371–387

Prudham, W. Scott. "Taming trees: Capital, science, and nature in Pacific slope tree improvement." *Annals of the Association of American Geographers* 93, no. 3 (2003): 636–656

Prudham, W. Scott. *Knock on Wood: Nature as Commodity in Douglas-Fir Country*. New York: Routledge, 2005

Rabinbach, Anson. *The Human Motor: Energy, Fatigue, and the Origins of Modernity*. Los Angeles: University of California Press, 1992

Raman, Sujatha, and Alison Mohr. "Biofuels and the role of space in sustainable innovation journeys." *Journal of Cleaner Production* 65 (2014): 224–233

Randalls, Samuel. "History of the 2°C climate target." *Wiley Interdisciplinary Reviews Climate Change* 1, no. 4 (2010): 598–605

Randerson, James. "Food crisis will take hold before climate change, warns chief scientist." *The Guardian*, 7 March 2008. https://www.theguardian.com/science/2008/mar/07/scienceofclimatechange.food (accessed 8 March 2025)

Rayner, Steve. "What might Evans-Pritchard have made of two degrees?" *Anthropology Today* 32, no. 4 (2016): 1–2

Realizing Increased Photosynthetic Efficiency. "Scientists engineer shortcut for photosynthetic glitch, boost crop growth by 40 percent." 3 January 2019. https://ripe.illinois.edu/press/press-releases/scientists-engineer-shortcut-photosynthetic-glitch-boost-crop-growth-40 (accessed 15 March 2025)

Renewable Energy Policy Network. *Renewables 2024 Global Status Report: Energy Demand*. Paris: REN21, 2024. https://www.ren21.net/wp-content/uploads/2019/05/GSR2024_Demand_Full_Report.pdf (accessed 26 September 2024)

Renewable Fuels Agency. *The Gallagher Review of the Indirect Effects of Biofuels Production*. St Leonards-on-Sea: Renewable Fuels Agency, 2008

Renshaw, Jarrett, Stephanie Kelly, and Leah Douglas. "Biden administration split over ethanol's role in aviation fuel subsidy program." Reuters, 2 August 2023. https://www.reuters.com/sustainability/biden-administration-split-over-ethanols-role-aviation-fuel-subsidy-program-2023-08-02/ (accessed 31 January 2025)

Rentier, Gerrit, Herman Lelieveldt, and Gert Jan Kramer. "Varieties of coal-fired power phase out in Europe." *Energy Policy* 132 (2019): 620–632

Reuters. "Biden team sets out path for ethanol aviation fuel subsidies." 1 May 2024. https://www.reuters.com/sustainability/biden-team-sets-out-ethanols-path-aviation-fuel-subsidies-2024-04-30/ (accessed 31 January 2025)

Ricardo Energy and Environment. *Analysing the Potential of Bioenergy with Carbon Capture in the UK to 2050*. 1 August 2020. https://assets.publishing.service.gov.uk/government/uploads/system/uploads/attachment_data/file/911268/potential-of-bioenergy-with-carbon-capture.pdf (accessed 24 March 2025)

Richardson, Tanya, and Gisa Weszkalnys. "Introduction: Resource materialities." *Anthropological Quarterly* 87, no. 1 (2014): 5–30

Robbins, Martin. "Fuelling politics." *Nature* 474 (2011): S22–S24

Robertson, Morgan M., and Joel D. Wainwright. "The value of nature to the state." *Annals of the Association of American Geographers* 103, no. 4 (2013): 890–905

Roni, Mohammad S., Sudipta Chowdhury, Saleh Mamun *et al.* "Biomass co-firing technology with policies, challenges, and opportunities: A global review." *Renewable and Sustainable Energy Reviews* 78 (2017): 1089–1101

Royal Society. *Sustainable Biofuels: Prospects and Challenges*. London: The Royal Society, 2008

Royal Society. *Net Zero Aviation Fuels: Resource Requirements and Environmental Impacts*. London: The Royal Society, 2023

Royal Society and Royal Academy of Engineering. *Greenhouse Gas Removal*. London: The Royal Society, 2018

Sandbag. *Playing with Fire: An Assessment of Company Plans to Burn Biomass in EU Coal Power Stations*. London: Sandbag, 2019

Sawyer, Donald. "Climate change, biofuels and eco-social impacts in the Brazilian Amazon and Cerrado." *Philosophical Transactions of the Royal Society B* 363 (2008): 1747–1752

Scheidel, Arnim, and Alevgul H. Sorman. "Energy transitions and the global land rush: Ultimate drivers and persistent consequences." *Global Environmental Change* 22 (2012): 588–595

Schelhas, John, Sarah Hitchner, and J. Peter Brosius. "What family forest owners talk about when they talk about trees: Bioenergy and forest landscapes in the U.S. South." *Trees, Forests and People* 17, no. 100606 (2024): 1–7

Schlesinger, William H. "Are wood pellets a green fuel?" *Science* 359, no. 6382 (2018): 1328–1329

Scholten, Daniel, Morgan Bazilian, Indra Overland, and Kirsten Westphal. "The geopolitics of renewables: New board, new game." *Energy Policy* 138, no. 111059 (2020): 1–6

Science News. "A plant for power alcohol." *Science* 83, no. 2157 (1936): 10a

Scott, James C. *Seeing Like a State: How Certain Schemes to Improve the Human Condition Have Failed*. New Haven, CT: Yale University Press, 1998

Scott, Vivian, Stuart Gilfillan, Nils Markusson *et al.* "Last chance for carbon capture and storage." *Nature Climate Change* 3 (2013): 105–111

Searchinger, Timothy D., Tim Beringer, Bjart Holtsmark *et al.* "Europe's renewable energy directive poised to harm global forests." *Nature Communications* 9, no. 3741 (2018): 1–4
Searchinger, Timothy D., Tim Beringer, and Asa Strong. "Does the world have low-carbon bioenergy potential from the dedicated use of land?" *Energy Policy* 110 (2017): 434–446
Searchinger, Timothy D., Robert Edwards, Declan Mulligan *et al.* "Do biofuel policies seek to cut emissions by cutting food?" *Science* 347, no. 6229 (2015): 1420–1422
Searchinger, Timothy D., Steven P. Hamburg, Jerry Melillo *et al.* "Fixing a critical climate accounting error." *Science* 326 (2009): 527–528
Searchinger, Timothy D., Ralph Heimlich, R.A. Houghton *et al.* "Use of U.S. croplands for biofuels increases greenhouse gases through emissions from land-use change." *Science* 319 (2008): 1238–1240
Seddon, Nathalie, Alison Smith, Pete Smith *et al.* "Getting the message right on nature-based solutions to climate change." *Global Change Biology* 27 (2021): 1518–1546
Sedgewick, Augustine. *Coffeeland: A History*. London: Allen Lane, 2020
Sieferle, Rolf Peter. *The Subterranean Forest: Energy Systems and the Industrial Revolution*. Cambridge: White Horse Press, 2001
Silva-Castañeda, Laura, and Nathalie Trussart. "Sustainability standards and certification: Looking through the lens of Foucault's *dispositif*." *Global Networks* 16, no. 4 (2016): 490–510
Simard, Suzanne. *Finding the Mother Tree: Uncovering the Wisdom and Intelligence of the Forest*. London: Allen Lane, 2021
Simon, Frédéric. "Brussels rebuked over 'confusing' draft EU bioenergy rules." *Euractiv*, 11 May 2021. https://www.euractiv.com/section/climate-environment/news/brussels-rebuked-over-confusing-draft-eu-bioenergy-rules/ (accessed 21 March 2025)
Simpson, Mark, and Imre Szeman. "Impasse time." *The South Atlantic Quarterly* 120, no. 1 (2021): 77–89
Simpson, Thomas W., Andrew N. Sharpley, Robert W. Howarth *et al.* "The new gold rush: Fueling ethanol production while protecting water quality." *Journal of Environmental Quality* 37, no. 2 (2008): 318–324
Smith, Neil. "Nature as accumulation strategy." *Socialist Register* 43 (2007): 16–36
Smith, Neil. *Uneven Development: Nature, Capital and the Production of Space*, 3rd ed. New York: Blackwell, 2010
Smith, Pete, Steven J. Davis, Felix Creutzig *et al.* "Biophysical and economic limits to negative CO_2 emissions." *Nature Climate Change* 6 (2016): 42–50
Smith, Pete, and John R. Porter. "Bioenergy in the IPCC assessments." *Global Change Biology Bioenergy* 10 (2018): 428–431
Smith, Stephen M., Oliver Geden, Matthew J. Gidden *et al.*, eds, *The State of Carbon Dioxide Removal 2024*, 2nd ed. https://doi.org/10.17605/OSF.IO/F85QJ
South, Paul F., Amanda P. Cavanagh, Helen W. Liu, and Donald R. Ort. "Synthetic glycolate metabolism pathways stimulate crop growth and productivity in the field." *Science* 363, no. 6422 (2019): 45
Southern Environmental Law Center. *Pellet Mill Community Impact Survey*, October 2024. https://www.selc.org/wp-content/uploads/2024/10/Biomass_Report_0924_F.pdf (accessed 17 March 2024)
Stephenson, Anna L., and David J.C. MacKay. *Life Cycle Impacts of Biomass in 2020: Scenarios for Assessing the Greenhouse Gas Impacts of Energy*

Input Requirements of Using North American Woody Biomass for Electricity Generation in the UK. July. London: UK Department for Energy and Climate Change, 2014. https://assets.publishing.service.gov.uk/media/5a7da114e5274a676d5332ff/BEAC_Report_290814.pdf (accessed 17 March 2025)

Sterman, John, William Moomaw, Juliette N. Rooney-Varga, and Lori Siegel. "Does wood bioenergy help or harm the climate?" *Bulletin of the Atomic Scientists* 78, no. 3 (2022): 128–138

Sterman, John, Lori Siegel, and Juliette N. Rooney-Varga. "Does replacing coal with wood lower CO_2 emissions? Dynamic lifecycle analysis of wood bioenergy." *Environmental Research Letters* 13, no. 015007 (2018): 1–10

Stoekl, Allan. *Bataille's Peak: Energy, Religion, and Postsustainability*. Minneapolis: University of Minnesota Press, 2007

Swinbank, Alan, and Carsten Daugbjerg. "The 2003 CAP reform: Accommodating WTO pressures." *Comparative European Politics* 4 (2006): 47–64

Tyndall, John. *Heat Considered as a Mode of Motion*. New York: D. Appleton and Company, 1869

Tyner, W.E. "Policy update: Why the push for drop-in biofuels?" *Biofuels* 1, no. 6 (2010): 813–814

UK Department for Transport. *Jet Zero Strategy: Delivering Net Zero Aviation by 2050*. London: UK Department for Transport, 2022

UK Department of Trade and Industry. *Our Energy Future—Creating a Low Carbon Economy*. London: The Stationery Office, 2003

Underwood, Ted. *The Work of the Sun: Literature, Science, and Political Economy, 1760–1860*. New York: Palgrave Macmillan, 2005

US Department of Agriculture Foreign Agricultural Service. *Global Agricultural Information Network Report: UK Wood Pellets Annual 2024*. https://apps.fas.usda.gov/newgainapi/api/Report/DownloadReportByFileName?fileName=Wood%20Pellets%20Annual_London_United%20Kingdom_UK2024-0033.pdf (accessed 17 March 2025)

US Department of Energy. *2016 Billion-Ton Report: Advancing Domestic Resources for a Thriving Bioeconomy*. July 2016. https://www.energy.gov/eere/bioenergy/articles/2016-billion-ton-report-advancing-domestic-resources-thriving-bioeconomy (accessed 16 March 2025)

US Department of Energy. "U.S. Department of Energy announces $61 million for biofuels research to reduce transportation emissions." 8 April 2021. https://www.energy.gov/eere/bioenergy/articles/us-department-energy-announces-61-million-biofuels-research-reduce (accessed 5 February 2025)

US Department of Energy. "DOE announces $590 million to increase bioenergy research." 17 March 2023. https://www.energy.gov/articles/doe-announces-590-million-increase-bioenergy-research (accessed 5 February 2025)

US Department of Energy. *2023 Billion-Ton Report: An Assessment of U.S. Renewable Carbon Resources*. March 2024. https://www.energy.gov/sites/default/files/2024-03/beto-2023-billion-ton-report_2.pdf (accessed 16 March 2025)

US Department of Energy. "Bioenergy KDF [Knowledge Discovery Framework]." https://bioenergykdf.ornl.gov (accessed 14 December 2025)

US Department of Energy. "Bioenergy Research Centers." https://www.genomicscience.energy.gov/bioenergy-research-centers/ (accessed 5 February 2025)

US Department of Energy, US Department of Transportation, and US Department of Agriculture. *Memorandum of Understanding: Sustainable Aviation Fuel Grand*

Challenge. September 2021. https://www.energy.gov/sites/default/files/2021-09/S1-Signed-SAF-MOU-9-08-21_0.pdf (accessed 16 March 2025)

US Department of Energy, US Department of Transportation, US Department of Agriculture, and US Environmental Protection Agency. *SAF Grand Challenge Roadmap: Flight Plan for Sustainable Aviation Fuel*. September 2022. https://www.energy.gov/sites/default/files/2022-09/beto-saf-gc-roadmap-report-sept-2022.pdf (accessed 16 March 2025)

US Federal Aviation Administration. *United States 2021 Aviation Climate Action Plan*. https://www.faa.gov/sites/faa.gov/files/2021-11/Aviation_Climate_Action_Plan.pdf (accessed 16 March 2025)

van der Horst, Dan, and James Evans. "Carbon claims and energy landscapes: Exploring the political ecology of biomass." *Landscape Research* 35, no. 2 (2010): 173–193

van der Wal, Sanne. *Wood Pellet Damage: How Dutch Government Subsidies for Estonian Biomass Aggravate Biodiversity and Climate Crisis*. Amsterdam: Centre for Research on Multinational Corporations, 2021. https://www.somo.nl/wood-pellet-damage/ (accessed 4 February 2025)

van Veelen, Bregje. "Negotiating energy democracy in practice: Governance processes in community energy project." *Environmental Politics* 27, no. 4 (2018): 644–665

van Vuuren, Detlef P., Andries F. Hof, Mariësse A.E. van Sluisveld, and Keywan Riahi. "Open discussion of negative emissions is urgently needed." *Nature Energy* 2 (2017): 902–904

Varinsky, Dana. "An energy company just debuted a shoe made from power plant emissions." *Business Insider*, 16 September 2016. https://www.businessinsider.com/repurposed-co2-shoe-without-footprint-2016-9 (accessed 24 March 2025)

Vermeulen, Sonja, and Lorenzo Cotula. "Over the heads of local people: Consultation, consent, and recompense in large-scale land deals for biofuel projects in Africa." *Journal of Peasant Studies* 37, no. 4 (2010): 899–916

Vickers, Noah. "Clearing the air on how wood pellet factories affect their neighbors and the climate." *The Appalachian Voice*, 28 January 2023. https://appvoices.org/2023/01/28/wood-pellet-nc-air/ (accessed 19 March 2025)

Virgin Atlantic. "Flight 100." https://corporate.virginatlantic.com/gb/en/business-for-good/planet/fuel/flight100.html (accessed 4 February 2025)

Vivid Economics. *Capturing Carbon at Drax: Delivering Jobs, Clean Growth and Levelling Up the Humber*. London: Vivid Economics Limited, 2020

Voegele, Erin. "USDA: US wood pellet exports top 926,024 metric tons in May." *Biomass Magazine*, 9 July 2024. https://biomassmagazine.com/articles/usda-us-wood-pellet-exports-top-926024-metric-tons-in-may (accessed 19 March 2025)

Walker, Gordon. "Keeping the lights on." In *Energy Fables: Challenging Ideas in the Energy Sector*, edited by Jenny Rinkinen, Elizabeth Shove, and Jacopo Torriti, 68–77. London: Routledge, 2019

Waller, Laurie, Tim Rayner, Jason Chilvers *et al.* "Contested framings of greenhouse gas removal and its feasibility: Social and political dimensions." *Wiley Interdisciplinary Reviews Climate Change* 11, no. e649 (2020): 1–17

Wear, David N., and John G. Greis, eds, *The Southern Forest Futures Project*. General Technical Report SRS-178, August. Washington, DC: United States Department of Agriculture Forest Service, 2013

Weeks, Kathi. *The Problem with Work: Feminism, Marxism, Antiwork Politics, and Postwork Imaginaries*. Durham, NC: Duke University Press, 2011

White, Samuel. "Forest owners' boss: 'Sustainable forest management crucial to maintaining carbon cycle'." *Euractiv*, 31 January 2018. https://www.euractiv.com/section/agriculture-food/interview/forest-owners-boss-sustainable-forest-management-crucial-to-maintaining-carbon-cycle/ (accessed 22 March 2025)

White House, The. "President signs Energy Policy Act." 8 August 2005. https://georgewbush-whitehouse.archives.gov/news/releases/2005/08/20050808-6.html (accessed 15 September 2024)

White House, The. "President Bush delivers State of the Union address." 23 January 2007. https://georgewbush-whitehouse.archives.gov/news/releases/2007/01/20070123-2.html (accessed 22 September 2024)

White House, The. "Executive order on advancing biotechnology and biomanufacturing innovation for a sustainable, safe, and secure American bioeconomy." 12 September 2022. https://bidenwhitehouse.archives.gov/briefing-room/presidential-actions/2022/09/12/executive-order-on-advancing-biotechnology-and-biomanufacturing-innovation-for-a-sustainable-safe-and-secure-american-bioeconomy/ (accessed 15 March 2025)

White House, The. "Fact sheet: President Biden to launch a National Biotechnology and Biomanufacturing Initiative." 12 September 2022. https://bidenwhitehouse.archives.gov/briefing-room/statements-releases/2022/09/12/fact-sheet-president-biden-to-launch-a-national-biotechnology-and-biomanufacturing-initiative/ (accessed 15 March 2025)

White House, The. *Bold Goals for U.S. Biotechnology and Biomanufacturing: Harnessing Research and Development to Further Societal Goals*. March 2023. https://bidenwhitehouse.archives.gov/wp-content/uploads/2023/03/Bold-Goals-for-U.S.-Biotechnology-and-Biomanufacturing-Harnessing-Research-and-Development-To-Further-Societal-Goals-FINAL.pdf (accessed 13 December 2025)

White House, The. "Remarks by President Biden on helping workers and innovators invent and make more in America." 28 July 2023. https://bidenwhitehouse.archives.gov/briefing-room/speeches-remarks/2023/07/28/remarks-by-president-biden-on-helping-workers-and-innovators-invent-and-make-more-in-america-auburn-me/ (accessed 15 March 2025)

Whyte, Kyle. "Indigenous climate change studies: Indigenizing futures, decolonizing the anthropocene." *English Language Notes* 55, nos. 1–2 (2017): 153–162

Wilkinson, John, and Selena Herrera. "Biofuels in Brazil: Debates and impacts." *Journal of Peasant Studies* 37, no. 4 (2010): 749–768

Williams, Michael. *Americans & Their Forests: A Historical Geography*. Cambridge: Cambridge University Press, 1989

Williams, Michael. "Dark ages and dark areas: Global deforestation in the deep past." *Journal of Historical Geography* 26, no. 1 (2000): 28–46

Winickoff, David E., and Matthieu Mondou. "The problem of epistemic jurisdiction in global governance: The case of sustainability standards for biofuels." *Social Studies of Science* 47, no. 1 (2017): 7–32

Wolford, Wendy, Ben White, Ian Scoones, Ruth Hall, Marc Edelman, and Saturnino M. Borras. "Global land deals: What has been done, what has changed, and what's next?" *Journal of Peasant Studies* (2024): 1–38. https://doi.org/10.1080/03066150.2024.2325685

Worldwatch Institute. *Biofuels for Transport: Global Potential and Implications for Sustainable Energy and Agriculture*. London: Earthscan, 2007

Worldwatch Institute. *U.S. Biofuels: Climate Change and Policies*. Washington, DC: Worldwatch Institute, 2009

Wrigley, Edward Anthony. *Continuity, Chance and Change: The Character of the Industrial Revolution in England*. Cambridge: Cambridge University Press, 1988

Yenneti, Komali, Rosie Day, and Oleg Golubchikov. "Spatial justice and the land politics of renewables: Dispossessing vulnerable communities through solar energy mega-projects." *Geoforum* 76 (2019): 90–99

Zale, Janice, Je Hyeong Jung, Jae Yoon Kim *et al.* "Metabolic engineering of sugarcane to accumulate energy-dense triacylglycerols in vegetative biomass." *Plant Biotechnology Journal* 14, no. 2 (2016): 661–669

Zero Carbon Humber. *Capture for Growth—A Roadmap for the World's First Zero Carbon Industrial Cluster*. Hull, UK: Zero Carbon Humber, 2019

Zoghlami, Aya, and Gabriel Paës. "Lignocellulosic biomass: Understanding recalcitrance and predicting hydrolysis." *Frontiers in Chemistry* 7 (2019): 1–11

Zwier, Jochem, and Vincent Blok. "Energetic ethics: Georges Bataille in the anthropocene." In *Global Changes: Ethics, Politics and Environment in the Contemporary Technological World*, edited by Luca Valera and Juan Carlos Castilla, 171–180. Cham, Switzerland: Springer Nature, 2020

Zwier, Jochem, Vincent Blok, Pieter Lemmens, and Robert-Jan Geerts. "The ideal of a zero-waste humanity: Philosophical reflections on the demand for a bio-based economy." *Journal of Agricultural and Environmental Ethics* 28 (2015): 353–374

Index

EU authorised representative for GPSR:
Easy Access System Europe, Mustamäe tee 50,
10621 Tallinn, Estonia
gpsr.requests@easproject.com

www.ingramcontent.com/pod-product-compliance
Ingram Content Group UK Ltd.
Pitfield, Milton Keynes, MK11 3LW, UK
UKHW061922210826
8081IPUK00002B/18
* 9 7 8 1 5 2 6 1 9 2 1 2 7 *

Insect Flight